Creating a Fire-Safe Community
A Guide for Fire Safety Educators

Creating a Fire-Safe Community

A Guide for Fire Safety Educators

Tom Kiurski

Fire Engineering®

Disclaimer
The recommendations, advice, descriptions, and methods in this book are presented solely for educational purposes. The author and publisher assume no liability whatsoever for any loss or damage that results from the use of any of the material in this book. Use of the material in this book is solely at the risk of the user.

Copyright © 1999 by Fire Engineering Books & Videos,
a Division of PennWell Corporation.

Library of Congress Cataloging-in-Publication Data
Kiurski, Tom.
   Creating a fire-safe community: a guide for fire safety educators / Tom Kiurski.
      p. cm.
   ISBN 0-912212-82-9 (softcover)
   1. Fire prevention--Study and teaching. 2. Community education.
I. Title.

TH9120.K57 1999
628.9'22'071--dc21
                                                                  99-052413

All rights reserved. No part of this work covered by the copyright hereon may be reproduced or used in any form by any means—graphic, electronic, or mechanical, including photocopying, recording, taping, or information storage and retrieval systems—without prior written permission of the publisher.

Published by Fire Engineering Books & Videos
A Division of PennWell Corporation,
Park 80 West, Plaza 2
Saddle Brook, NJ 07663

Edited by Mary Jane Dittmar
Book layout by Brigitte Pumford-Coffman
Cover design by Brian Firth

Printed in the United States of America

99 00 01 02 03   1 2 3 4 5

To Nancy

Acknowledgments

Firefighters need as much help as they can get in preparing for a much-needed and often overlooked aspect of their job: fire safety education. For years, I have been searching for a simple, easy-to-follow manual to assist in the development and delivery of such a program. Finding none, I developed this manual in the expectation that it would make teaching fire safety easier. I acknowledge the contributions of many individuals who have helped with this project. My wife and best friend, Nancy, has been by my side during my fire service career. Our children, Stephen and Beth, have helped me in presenting programs on more than one occasion.

Other talented family members have helped in many ways. Special thanks go to my sister-in-law Gayle Kiurski and my brother Ken for their numerous editorial suggestions.

My friend Bill Anderson provided his computer services and artistic talents. Gayle Gerig and Cindy David from the Wayne County Regional Educational Services Agency, Regional Education Media and Technology Center (REMTEC), Livonia Satellite, have done so much over the years to make my programs run smoothly. Numerous local school principals, teachers, administrators, and other staff members have helped beyond my ability to express.

I also recognize the contributions of all who support fire safety education. Whether they deliver fire safety messages, help with tours, or fill in for others, all of them help the cause.

My home team, the Livonia firefighters, have given their all when needed, and they're always there to help educate our citizens. Listening to ideas, offering suggestions, building projects, and giving up their own free time to help advance fire safety education—these firefighters have done it all. Space doesn't permit me to list all of their names, but their hard work and perseverance are deeply appreciated.

Table of Contents

Introduction 1

Chapter 1: Planning a Public Education Presentation 3

Chapter 2: Basic Fire Safety Concepts 11

Chapter 3: Schools 27

Chapter 4: Scouts 51

Chapter 5: Water Safety 65

Chapter 6: Industrial Safety 71

Chapter 7: Medical Emergencies 79

Chapter 8: Evacuation Drills 85

Chapter 9: Adult Education 97

Chapter 10: Seasonal Safety Programs 103

Chapter 11: Preventing a Poisoning 125

Chapter 12: Public Service Announcements 129

Chapter 13: Print Messages 147

Chapter 14: Props, Gifts, and Books 153

Chapter 15: Games and Tests 163

Chapter 16: Sample Handouts 175

Introduction

Many of today's fire safety educators have gotten started on the wrong foot. The topic of teaching audiences is one that may have been handled in a cursory way in recruit school. It isn't very macho to teach public education programs. The instructor may have been better at forcible entry than at methods of education. Someone may have even revealed the unspoken contract between firefighters and instructors: "They teach and we fight fires, and we try not to impinge on each others' turf."

Whether or not your instructors slighted the subject, the information in this book will enable you to start new fire safety education programs or to improve existing ones in a short period of time. No additional books, tapes, or computer disks are necessary. You will be ready to make an effective presentation within a few minutes of being invited to do so.

A favorite story around fire stations is one in which a firefighter walks out of the building, saying, "I'm going out on an area inspection," then learns as he arrives at a local school that he's expected to present an education program for the students. Although last-minute prepping isn't the best way to make an educational presentation, the information contained in this book can, in a crunch, help you assemble an impromptu fire safety lesson. It might be a good idea to keep a copy on the apparatus for that purpose. A teacher or an assistant can prepare handouts while you are conducting the lesson.

Firefighters aren't expected to fight fires without training, nor could they perform emergency medical services without certification, yet in some departments, they are sent out into the community to teach fire safety without the least bit of educational training. We all know how to entertain grade-school kids by demonstrating our gear and squirting water into a field. Such activities may make an impression, but they teach nothing. Firefighters may be dedicated to responding to emergencies, yet, as a group, we choose not to teach fire prevention and safety. If our mission is to protect the community, then our overall strategy must include education as well as suppression.

Teaching fire safety can be fun, and squirting water can still be done at the proper time. Our main task in this respect, however, is to educate the citizens.

Many teachers do a good job of presenting a fire safety curriculum, but how many of them have witnessed the effects of mushrooming smoke and heat? Do they know how many fires can easily be prevented or how rapidly flames can spread? Do they know about the toxicity of many modern combustibles and how crucial it is to have a plan for quickly and safely escaping from a fire?

This book isn't intended to take the place of formal training, but it can help you inaugurate an effective program with a minimum of research and preparation. I encourage you to enroll in the public fire safety education and instructional methodology programs offered by the National Fire Academy (NFA) and to pursue National Fire Protection Association (NFPA) certification to the level of Fire Prevention Education Officer. In the meantime, however, if you must prepare a program that is scheduled for tomorrow or the day after, the information contained herein can make it easier, and even fun, to teach fire safety.

Chapter 1

Planning a Public Education Presentation

Preparation

As stated by the International Fire Service Training Association (IFSTA), developing a public education presentation involves four steps: preparation, presentation, application, and evaluation. Once the idea to present public education has taken root, the first working step toward it is in preparation. Who will be in the audience and what topics should be included in the presentation are primary considerations when preparing your program.

Audience

The nature of the audience is usually predetermined, as in those instances when you are invited to speak at a school, a service club luncheon, or the meeting of a tenants association. Other times, your department may set up a schedule that will enable you to reach certain population segments with specific messages.

In any case, try to learn as much about your audience as possible, such as the estimated number of attendees, the age groups represented, and any other information that may indicate topics of special interest or need. Research will guide you. A review of geographical areas, for example, might reveal that one segment of the population experiences a higher incidence of fires than the rest. You may want to develop a program that highlights the problem and ways to reduce the risk, then present it at schools in that neighborhood, before local organizations, or at a neighborhood-at-large meeting.

Your department may want to hire a professional market research firm to make an in-depth study of the target group's attitudes, perceptions, misconceptions, and behaviors with respect to fire-related issues. Consultants function as sounding boards for the fire department by gaining access to and analyzing information that might otherwise never be brought to the attention of fire department managers and officers. Consultants can also help you determine which segments of the target audience can best assist in getting out the message.

Content

The content of your presentation should meet both the audience's needs and interests. Sometimes your research will help you select both your audience and the presentation topics, as in the case of the high-risk neighborhood mentioned above. Other times, the problem that needs addressing may be one that is of major concern in your overall jurisdiction. By reviewing fire reports, you can identify the major causes of the fires that have occurred in your community. In your presentation, you would point out the causes of those fires and the actions that could help prevent them and promote fire safety.

Establish and write down measurable objectives that can be used to evaluate the effect, if any, of the educational program. Suppose your basic message was that smoke detectors in homes save lives. Your objective might be to spur an increase, perhaps targeting a specific percentage, in the number of homes in which detectors are installed. First, you would have to determine how many homes already had detectors at the time you instituted your program. You can obtain this information through a random telephone or mail survey. That same sample can be used for a follow-up survey within a specified time after your presentation.

Appearance

Preparation includes appearance. Research has shown that a speaker's appearance can greatly affect his credibility. Get out your cleanest uniform, and look in the mirror before you leave your house or the fire station.

Check Details

Being prepared also means arriving on time and having all of the equipment and supplies that you'll need for the presentation. Be sure you know the name and address of the meeting place and the time that you're expected to arrive. Determine what facilities and equipment will be available, including anything that you may need in the way of audiovisual equipment, copy machines, and the like. Be sure to get the name and telephone number of a contact person in case you have questions or an emergency arises.

Verify in advance approximately how big your audience will be so that you'll know how many handouts to prepare. If the presentation is to be given at the fire station, you'll have to arrange for adequate seating and possibly even refreshments.

Be well prepared. Plan your presentation, practice your delivery, and have all handout and reference materials ready for distribution.

Presentation

Presentation involves tailoring the method of instruction to the subject matter and the audience. The message you intend to convey may best be presented by means of a lecture, a demonstration, a group discussion, audience participation and role-playing, or through some other format. If possible, test your program with a segment of the target audience and make any necessary adjustments. Maintain eye contact with your audience as you speak. If direct eye contact makes you nervous at first, focus on proximate objects, such as a person's tie, his water glass, or a picture on the back wall. Adopt a relaxed, confident manner. Standing at attention can be uncomfortable for you and your audience. Use hand gestures to emphasize key points.

Remember to acknowledge and thank the volunteers and community leaders who helped make the event possible and who assisted with the preparations.

Techniques

Introduce the subject matter with enthusiasm and creativity to motivate your audience. Speak at their level of understanding. Relate real-life problems and incidents, and stress the importance of their learning the information that you're presenting. Encourage audience participation, and involve everyone. Reinforce your message with teaching aids that you prepared well in advance, such as films, slides, handouts, videos, overhead transparencies, and props. For example, demonstrate how even a little smoke can activate a smoke detector, or show a detector taken from an actual fire building, pointing out the melted area around the battery and other characteristics.

Distribute handout material at the end of the program. The participants will be able to review it at home and share it with family members. Encourage the participants to give you some honest feedback. Ask what you might do to improve the program.

Application

This step involves enforcing your message so that those who attended your presentation will be prepared to apply the principles they learned. The follow-up program may include public service announcements on local TV stations, print ads, or feature articles on the use of smoke detectors. The enforcing mechanism might consist of reminder stickers that a citizen can place on his calendar, refrigerator, or elsewhere to indicate when home detector batteries should be tested.

Advertise, hold press conferences, issue press releases, and promote other activities that will call attention to your program. Make the public aware of incidents that demonstrate how the information given in the program has benefited citizens, such as by saving a life or property. (The individuals involved in such incidents may have to give their permission.) Publicize activities relating to the program's goals.

Evaluation

You can measure the effectiveness of any given program in various ways. One short-term mechanism is to have the audience take a test before the presentation and then another one afterward, so as to measure what they learned. One example of a long-term evaluation device is a survey. As mentioned above, you might undertake a survey to determine how many families have installed smoke detectors in their homes and have kept them in working order since the implementation of the program. Or, you might note how statistics relating to fires, deaths, and injuries have changed within a specific neighborhood or the overall community.

You could also track the change in the number of fires that occur in the type of structure targeted in the program. For example, if you were evaluating cooking-related fires, you might find that they have gradually declined over a three-year period.

You may not meet your goals, but don't be disappointed. Teaching the public to be more fire-safe is a great service to render. Don't get discouraged. You can always adjust the program to help your department achieve the desired results. The adjustment may necessitate adopting an evaluation instrument that measures the targeted variables more closely, or you may simply have to extend the time frame for reaching the objective.

Chapter 2

Basic Fire Safety Concepts

Sticking to Basics

Most instructive programs of any kind boil down to making an audience aware of a few key points. A rule of thumb in virtually any kind of education is that the more elementary a skill or given bit of knowledge is, the more valuable it is. A general, fundamental rule can be more generally applied in everyday life than one that is tied to more advanced principles. For the fire service, this means educating an audience on basic means of prevention and coping with emergencies—subjects that professional firefighters might almost take for granted or consider obvious, but about which the average citizen may never have given much thought.

This chapter is comprised of some of the more common subject areas of fire prevention and the key points associated with them. The points listed under each heading, presented in an outline format, serve as the basic backbone for a presentation on each subject.

Matches and Lighters

- Matches and lighters are tools.
- There are good and bad fires. Good fires include those in fireplaces and

barbecue grills. Bad fires are those that are out of control, cause destruction, and kill people.
- Matches and lighters can start both friendly and unfriendly fires.
- Only grown-ups should use these tools.
- Children can be fire helpers by blowing out the candles after dinner, by gathering wood for campfires, by helping to carry cooking tools and food to the grill, and by bringing in wood and paper for the fireplace.
- Some grown-ups get careless with matches and lighters and may leave them where children can get ahold of them. If you should see these tools anywhere but in a safe place, tell a grown-up, whether your parents, an older relative, or your baby-sitter. Never touch these tools or give them to a friend.
- When you get older, you will learn how to use these tools correctly to start friendly fires.

Frequent Causes of Fire

- The main causes of fire are heating equipment, unsafe cooking habits, electricity, smoking, and arson.
- All heating sources, including hot water heaters, radiators, furnaces, and portable space heaters, need a three-foot clearance on all sides.
- Change furnace filters regularly, per the manufacturer's recommendation or that of a plumbing-heating professional.
- Have furnaces inspected regularly, also per professional recommendation.
- Keep children away from heating equipment.
- Never leave the room when a space heater is on.
- Use the appropriate fuel in fuel-operated engines.
- After lighting the fireplace, cover the opening of the fireplace with a screen.
- A responsible adult should attend a fire in a fireplace.
- Make sure that combustibles are at least three feet away from the hearth.
- Do not allow horseplay near the fireplace.

Cooking Safety

- The kitchen is the most dangerous room in the house. Most cooking-related fires are caused by leaving food unattended on the stove. Cooking

should always be supervised. If you must leave the kitchen while cooking, shut off the burners. Otherwise, take a utensil or a pot holder with you to remind you to return to the kitchen.
- Turn pot handles inward toward the stove so that young children cannot grab them.
- Cook on the back burners, away from young children.
- When cooking, don't wear loose clothing, such as bathrobes and dresses with dangling sleeves.
- Have a set of large oven mitts nearby. Cooking mitts are safer than pot holders. They protect better against burns.
- Do not use water against a food-on-the-stove fire. The steam that's produced can scald you, and the splatter can actually contribute to the generation and spread of flames.
- Keep a lid nearby that fits the pan in which you are cooking. In case of a grease fire, slide the lid over the pan with a gloved hand and turn off the heat. Keep the lid in place even after the flames die down. If you remove it too soon, the fire might flare up again.
- You can also put out a pan fire with an extinguisher rated for Class B fires. Know where the fire extinguishers are in your home.
- Pour large quantities of baking soda on the fire. Never try to carry the burning pan outside. The air can push the heat toward you, and you may drop the burning pan on the floor and cause the fire to spread.
- If the fire is in the oven, shut off the oven and keep the oven door closed. Keeping the door closed will confine the heat and limit oxygen, thereby starving the fire and preventing extension.
- Do not overfill pots. An overfilled pot may boil over.
- Clean cooking equipment regularly so that grease doesn't build up under the pans. Grease can make a cooking fire grow faster and spread farther.
- If a fire occurs, notify the fire department without delay.

Smoking Materials

- Smoking materials are involved in the greatest number of fire deaths.
- If you or someone in your home smokes, be extremely careful. Make

it a rule that no one smoke when he is tired or in bed. If you are drowsy, extinguish all smoking materials.
- Use large ashtrays with high sides.
- Empty ashtrays in the toilet. If you do dispose of butts in the garbage, make sure they are extinguished and the ashes are cold.
- Don't leave smoking materials unattended.

Electrical Hazards

- Inspect electrical cords at least annually to be sure they're in good working order.
- Do not run extension cords under rugs.
- Never use extension cords as permanent wiring.
- Limit the number of appliances that you plug into a single outlet. Consider installing ground-fault circuit interrupters (GFCI) in your home. These outlets are equipped with a miniature circuit breaker that will cut off the electricity to the outlet if a short occurs.
- Use UL-approved appliances. Where several cords are plugged into the same outlet, consider using a UL-approved outlet strip with a built-in circuit breaker.

Fire Triangle and Tetrahedron

- A fire triangle is a way of expressing in graphic form the three components needed for fire to exist: fuel, oxygen, and heat.
- The fuel may be in any of the three conventional states of matter: solid, liquid, or gas. All fuels must vaporize to burn, since only vapors burn.
- The atmosphere consists of approximately 21 percent oxygen. A fire needs at least 16 percent oxygen to burn. Oxygen itself doesn't burn. It supports combustion by combining with the fuel (oxidation) in the presence of heat to form new compounds.
- For a material to burn, it must be raised to the proper temperature. The source of this heat may be by any means, whether chemical, electrical, mechanical, nuclear, solar, or otherwise.

- The fire tetrahedron represents in graphic form all of the components of the fire triangle, plus the additional component of a chemical chain reaction, in which a series of molecular interactions serve the combustion process.

Smoke Detectors

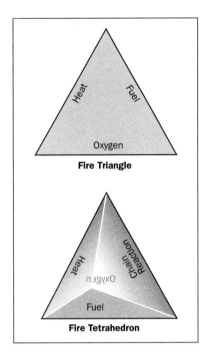

- Smoke can make you very sick. A smoke detector can smell smoke. It can warn you of fire and give you time to get out of the house before the fire gets too large.
- Having a working smoke detector in your home can cut by one-half your risk of death in a house fire. There is at least one smoke detector in about ninety percent of the homes in this country, but almost a third of them don't work.
- Your home should have several smoke detectors, and all family members should know what they sound like when they go off. Family members should plan in advance what to do if the alarm goes off.
- There should be at least one smoke detector per floor. Ideally, there should be one in every room. At minimum, there should be one in the basement, as well as one outside and one inside of each bedroom.
- Since you should be sleeping behind closed doors (to prevent fire spread), the warning of a smoke detector will allow you to escape from a fire before open flames break out. Without this early warning, you might be overcome by carbon monoxide fumes.
- Do not place detectors in areas that are prone to false alarms, such as the cooking area of the kitchen and near fireplaces.
- Once a month, test the detectors to see whether they're working properly. Use a broom handle to push the button so you won't have to climb on a ladder or a chair. This is particularly important for senior citizens. Dust the

detectors regularly and change the batteries once a year. Fire Prevention Week in early October is a good time to install fresh batteries.

Fire Extinguishers

- Many homeowner insurance policies offer reduced rates if the home is equipped with fire extinguishers. Sprinkler systems, smoke detectors, and carbon monoxide detectors are other fire-related aids that can help lower insurance rates.
- Purchase only multipurpose or ABC-type fire extinguishers for your home. The letter designations refer to different types of fires, all of which can occur in a home. Class A fires involve ordinary combustibles. Class B fires involve flammable liquids. Class C fires involve energized electrical equipment.
- Mount the fire extinguisher between the main hazard areas and an exit so that, if the extinguisher cannot control the fire, you'll have a quick, easy means of escape.
- In the workplace, know where the fire extinguishers are located and the classes of fire for which they are effective.
- In the event of fire, call the fire department before attempting to use an extinguisher.
- When using a fire extinguisher, remember the acronym PASS, which stands for pull, aim, squeeze, sweep. Pull the pin, aim at the base of the flame, squeeze the trigger, and sweep at the base of the flames from side to side.
- Most portable home fire extinguishers have only about ten seconds' worth of capacity, so quick action when the fire is still small is of paramount importance.

Call 911 for Emergencies

- 911 is a free call from a pay phone.
- If the number for emergency calls in your area isn't 911, make sure you know what the right number is.
- Use 911 only in case of a fire, serious injury, medical problem, or other

such emergency. Do not use it for a leaking pipe or for help in retrieving your keys out of a locked car.
- Speak calmly and clearly.
- Give the dispatcher as much accurate information as possible, including the address, nearby cross streets, the type of emergency, the number of people involved, and the extent and type of injuries.
- Remain on the line to answer the dispatcher's questions. It is only by giving accurate, detailed information that the appropriate number and type of vehicles can be sent to help.

Safety Reminders

- Close the doors to the bedrooms when you go to sleep. Closed doors can keep smoke and heat away long enough for you to use another exit. If you are trapped in a bedroom and must stay there until firefighters arrive, the closed door should protect you for a short period of time.
- If a fire occurs during the night, roll out of bed onto the floor. Do not stand up, or you might become disoriented by the poisons in the smoke. Always crawl low in smoke.
- When leaving the fire area, close the doors behind you to keep the fire from spreading.
- Never reenter a burning building for any reason.
- Stay outside to wait for the fire department. Tell the firefighters whether all of the family members are at the designated meeting place. If someone is missing, tell the firefighters where he or she was last seen.

Surviving a Fire

- Smoke and heat rise, so crawl low under smoke. The difference between crawling and walking could be the difference between life and death in a fire. The cooler, cleaner air is down low, near the floor. If you crawl, you'll be able to see and breathe better, but do not squirm along on your belly. Crawl on your hands and knees. If you're too close to the floor, heavier-than-air toxins, such as those that emanate from burning plastics and foam rubber, may harm you.

If Your Clothing Catches Fire

- If your clothing catches fire, stop what you're doing.
- Drop to your knees, right where you are. Then, fall flat on the ground with your legs out straight.
- Roll over in a back-and-forth manner while covering your face with the palms of your hands.
- Be sure to roll on all four sides—your front, back, left side, and right side. Young children often rock on two or three sides only.

Burns

- There are three categories of burns: first, second, and third degree.
- A first-degree burn looks like a sunburn. The skin is red. Flush a first-degree burn with cool water. This type of burn can heal itself. Keep the affected area in cool, clean water for five minutes. The water will take away some of the heat from the skin and prevent the skin from becoming more deeply burned. Do not use ice cubes or ointments.
- A second-degree burn isn't only red, it also has blisters and is painful. Do not pop the blisters. Cover a second-degree burn with sterile gauze or a sheet. Wrap a second-degree burn in a clean sheet and blankets. Call a doctor for this type of burn.
- A third-degree burn is grey or black in appearance. It affects the deep layers of the skin and the nerve endings. This type of burn needs immediate medical treatment.
- There is no such thing as a fourth-degree burn, although journalists sometimes use this term to describe third-degree burns that penetrate to the bone.
- In all cases, immediately put cool water on the burn for at least ten minutes.
- If your clothes are burning, stop, drop, and roll on all four sides to smother the fire. Then, cool the burn with water.
- If you're baby-sitting and the child suffers a burn, immediately call the child's parents or the grown-up in charge.

- If you're a minor, always tell your parents or another adult if you incur a burn. The burn may have to be treated by a doctor, especially if the skin is blackened, white, or blistered.
- Call 911 for treatment and transportation, if needed.
- While waiting for medical help to arrive, place a clean, dry bandage or cloth over the burn after cooling the area for ten minutes in water.
- Many burns are caused by exposure to scalding water or open flame. Prevent scalding by checking the temperature of the bathwater before getting in or placing a child into it. If the hot water in your home is consistently too hot, set the gauge on the water heater to a lower temperature.

Carbon Monoxide

- Carbon monoxide is a colorless, odorless, poisonous gas that can make you sick or kill you.
- A concentration of greater than 1.3 percent in air can kill you if you continue to breathe it for more than three minutes.
- Shorter exposures can make you dizzy, weak, and unable to think clearly.
- All fires produce carbon monoxide, and CO is the leading cause of death in fires. In fact, the carbon monoxide in smoke kills three times more people than burns caused by flames.
- Like heat and smoke, carbon monoxide rises within a building before it starts to bank down, so always stay low in a heat- and smoke-filled environment.
- Among other reasons, firefighters must extinguish fires quickly to minimize the amount of carbon monoxide that is produced.

Home Escape Plan

- Design a home escape plan, and have all of your family members sign it to ensure that they're all aware of it.
- Sketch the entire house, labeling all of the rooms, the primary and secondary routes from each room, and the designated outdoor meeting place.
- If the door to your bedroom isn't hot, keep your body weight pressed

against it and open it slightly. If there's little or no smoke present, quickly crawl out the nearest exit and go to the family meeting place.

- If the door to your bedroom is hot, don't open it. Use the secondary exit for escape and go to the family meeting place. If smoke and fire gases are coming in around the top of the door, or if you can see the orange glow of fire, use the secondary exit.
- If you choose a window as a secondary exit, be sure that you can open it. Also, know where the window leads.
- Note the placement and operational status of all smoke detectors.
- Take all of the family members on a tour of the house to identify the primary and secondary means of exit, and ensure that they all know how to use them in the event of a fire.
- Practice the plan with your family.
- Establish a plan for siblings who are too small to open windows.
- The outdoor meeting place should be close to the home but not attached to it.
- Review the actions to be taken at the family meeting place. Once two members are there, one can go to a neighbor's house or elsewhere to call the emergency number to notify the fire department.
- Select one family member to report to the first-arriving fire company, telling the officer the status of each family member.
- Assign a responsible person to be in charge of family members who cannot exit the house without help, such as infants and the handicapped.
- If you cannot remove a handicapped person from the house, lower him or her to the floor near a window. Open the window. Have the handicapped person hold a towel or piece of clothing over his nose and mouth and tell him to breathe through it. Then, leave him where he is, exit the house, and inform firefighters as to that person's location.
- Keep a whistle in every bedroom. If you're trapped, stay low, open the window, blow the whistle, and wave a sheet outside.
- If you're trapped in a room in which there is a telephone, dial 911 and tell the emergency dispatcher where you are. The dispatcher will inform the firefighters at the scene.
- If the fire department hasn't yet arrived, and if a family member is trapped, inform the dispatcher of this, giving the location of the room and as much other information as the dispatcher requests.

Basic Fire Safety Concepts

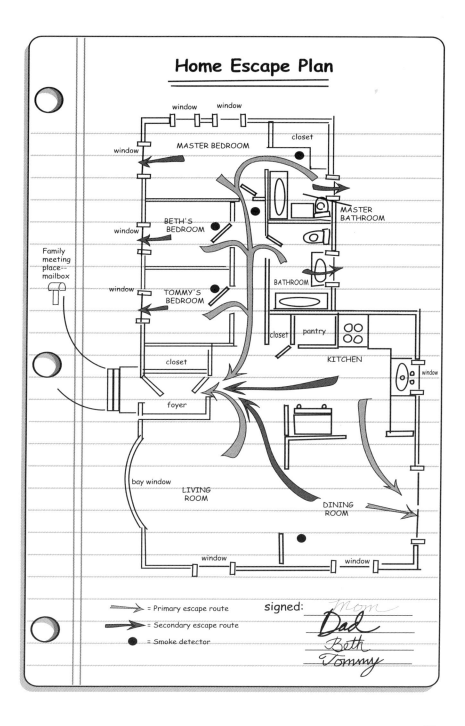

Multiple-Family Dwellings

- If your building has an elevator, never consider using it as an emergency exit.
- In a high-rise building, the interior exitways should be marked with exit signs.
- If the building has a pull alarm system, all of the members of your family should note the locations of the alarms and know how to activate them in case of an emergency.
- If there is an escape ladder outside the window, carefully use it to descend to the ground or the lowest point possible.
- If a window leads to a garage roof, and if the garage is a safer area than where you are, carefully climb out onto the roof and go to the lowest point.
- If you need to break a window, stand off to one side and strike the pane with a bat, stick, or some other tool. Then, poke the tool around the window frame to remove any remaining shards of glass. Place a quilt or bedspread over the sill to act as a protective layer before you go out.
- Use a commercially available portable escape ladder. You can store it under the bed and place it in the window, if needed.
- In lieu of a portable escape ladder, you may be able to climb down a knotted rope. Be sure the rope is securely anchored to the bed or some other large, heavy object. Tie a knot every six feet along the rope to act as a foothold.
- If you are forced to jump, you can reduce the fall distance by using the hang-drop method. Straddle the windowsill with one leg inside and one leg outside. Then, bring your inside leg across the sill and lower it until you are hanging by your outstretched arms, then drop to the ground.

Senior Citizens

- Keep a telephone, a whistle, and a pair of eyeglasses by your bed.
- Always put your glasses on for any emergency.
- Have a plan of escape, and practice that plan. The poisons in smoke can affect your judgment. Plan ahead so that you'll know what to do.
- Sleep with your bedroom door closed, since this will allow you extra time to escape in the event of fire.

- If there is a fire, leave your valuables behind, and call 911 from a neighbor's phone. Never reenter the house.
- If you cannot escape, call 911 and give them your location. Hang a sheet or blanket out the window, and blow the whistle.

Chapter 3

Schools

Lower Elementary School Programs

Young children make good audiences. Often they come up with wild comments and questions that'll make you laugh. I remember once asking a group of kindergartners what they would do if their clothes caught fire. One child raised his hand. I was hoping for the standard response, "Stop, drop, and roll." Instead, he surprised me by saying, "I wouldn't put them on!"

Students at the lower grade-school levels openly admire firefighters as heroes. Just listen to the first child who catches sight of you, announcing to his friends, "The firefighter is here! The firefighter is here!" Before your ego grows too big, however, you'd better be ready for their inevitable follow-up question: "Are you a real firefighter?"

When setting up a presentation for the lower grades, ask your contact whether other classes at the same level might participate. You may have only been invited to speak to one class, even though others might care to join in.

This may be the first time that the kids have met a firefighter in person. Even if they worship you, children at this age get fidgety quickly, so keep them actively involved in your presentation.

The Firefighter Is Your Friend

One of the basic messages that you should give to children in the lower grade levels is that a firefighter is their friend. Let the kids know that firefighters also respond to car accidents and medical emergencies. Firefighters teach people how to be fire-safe, and they inspect buildings to make sure that there are no hazards in them. And, of course, they put out fires.

Kids should know that they can go to firefighters for help in all sorts of emergencies. Tell them that sometimes firefighters wear work uniforms that cover their dress uniforms.

Put on your protective clothing one piece at a time while explaining what each component is and how it serves your safety. Barging in the door fully dressed in your protective gear may frighten young children, who may think that a monster has just burst into the room. (However, upper elementary students will appreciate such an entrance.)

When fully dressed in your protective gear, get down on your hands and knees, then move among the children. Let them see you close up, but don't frighten them. Show them how you would crawl through a burning building, probably in the company of a partner, searching for people in need of rescue. Show them how the hose brings clean air from the tank on your back to your face mask. Keep the terms simple. Let them become familiar with your appearance, and don't let anything about your protective gear seem threatening. Liken your appearance to a big, yellow elephant with a black trunk.

Reassure the children that the firefighter inside all that gear is still the same friendly person they saw before in his street clothes. Instruct them that they shouldn't run away, but that they should let the firefighter take them out of the hot, smoky building to the outside, where the air is clean and there is no danger from the fire.

As you crawl around among the children, offer your gloved hand to them to shake, or give them a high five. Friendly contact such as this will help them make the transition from seeing you as a strange, scary figure to someone who is trying to help them.

At the conclusion of your demonstration, stand up and take off your protective gear so they can see you turn into a human being again.

Call 911 for Emergencies

Ask the students what number they should dial to get firefighters in an emergency. Many will raise their hands. Kids often watch the show "Rescue 911" on television, so many of them are quite familiar with the number. If the emergency number in your area isn't 911, make sure they know what number to call.

When they give you the correct number in response to your question, act surprised. Ask a student to say the number, and praise him for giving the correct answer. Repeat the number so the entire audience can hear it. Then have all the students say it together. Pretend that you can't hear them, and ask them to repeat it louder. Punctuated repetitions such as these can help them commit the number to memory.

Use an old telephone or a toy phone as a prop. Hold it up while asking for the number that you should call. Move your fingers so the children can see which buttons you're pushing. Press 911.

As a prop, create the letters 911 out of foam rubber or laminated cardboard. Make them about three feet tall. Pass them out to three children. Have these three stand in front of the group with the numbers showing, although not in the correct order. Have the audience yell out to you what to do to make the numbers read 911. Switch the digits around a few times before you finally catch on to what they're saying.

Have the students practice making an emergency call.

Stop, Drop, and Roll

It is important that young children learn how to defend themselves if their clothes should suddenly catch fire. Make them realize that their clothing could catch fire in various ways. Describe how they might back into a barbecue grill or trip into a campfire, then ask them what they should do if this happens. Hopefully they'll come up with the phrase "Stop, drop, and roll." If they know what actions to take but don't know the phrase, tell them they know the right procedure, but that they should still remember the phrase. Make them repeat the words several times to reinforce their meaning.

As the hands go up, pick a volunteer helper to show the class how to perform the stop, drop, and roll. Select one of the bigger children; someone who raised his hand quickly. At such an age, many of the youngsters will put up their hands just because the others are doing so.

As the volunteer comes forward, ask the class whether they think that he or she knows how to stop, drop, and roll correctly. Then, explain that all a person has to do is to stop what he's doing, drop to his knees right where he is, and roll over in a back-and-forth manner while covering his face with the palms of his hands.

While explaining the basic movements to the class, you're also telling the volunteer what he should do, thus giving him confidence to demonstrate the technique.

Have the entire class count to three and then have the volunteer perform the series of movements. Praise his performance. Try to get him to roll over about four times before stopping to make sure the fire is out. Instruct him as necessary until he performs the movements correctly. Also, make the class say aloud the name of each movement as the volunteer performs them, since this will help reinforce the lesson. Depending on how much room is available, you may choose to have two or three children practice at a time or let the entire class practice together. If the audience is large, you may have one class compete against another. Make sure that there is enough room so that the children don't simply bump around in a tangle, and have the teachers assist you in maintaining decorum and reinforcing the proper technique.

Crawl Low Under Smoke

To help teach youngsters that they should crawl low under smoke, arrange some tables end to tend, and have the children crawl under the tables for their entire length. You might also have them crawl under a broom handle that has some dark-colored streamers hanging from it, or perhaps have two adults hold a blanket at waist level and let the children crawl underneath it. If you prefer to create a more permanent prop, you can build a fixed tunnel out of some PVC piping, elbows, and tees. Make the tunnel about six feet long. The two sides of it should be identical. Connect the sides with a few crossbars on the top, but leave several of the components unglued for quick disassembly. To make reassembly easier, you can color-code the pieces that join together. During the exercise, you can simulate dark, smoky fire conditions by placing a blanket over the top of the tunnel or by shutting off some lights. Explain to the kids that it's difficult to see through smoke and that smoke will make the inside of a building dark. One other way to construct a smoke tunnel is to put several sawhorses together. Have the children crawl under the horizontal member of each horse. If the sawhorses are wooden, use appropriate padding to prevent injuries.

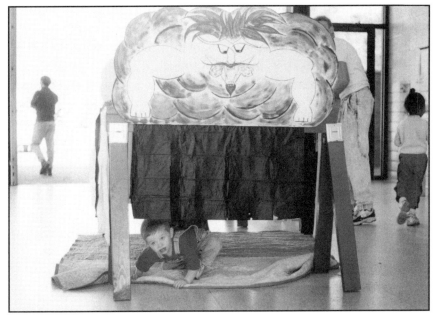

To help teach youngsters to crawl low under smoke, create an imaginative tunnel about six feet long.

Smoke Detectors

Bring a smoke detector to the presentation. Before showing it to the children, have them think of smells that are familiar to them. Bring up ideas such as chocolate chip cookies baking in the oven or flowers blooming in the spring. Then, mention the smell of a skunk as a lead-in to the smoke detector.

Reinforcing the Message

Present some pictures depicting various emergency scenarios, such as buildings filled with smoke or people wearing clothes that are starting to catch fire. The pictures may be photographs or they may be drawn by hand. Have the children identify the appropriate survival technique for each scenario, such as crawl low under smoke or stop, drop, and roll.

Run through the key parts of your presentation again. This time, however, leave out some key points. Make believe that you can't remember them, and have the students fill in these parts. Deputize the children, saying, "I need *you* to help teach people to be fire-safe. Can you help me to teach others?" Once they answer yes, request that they teach their family members what they've learned about fire safety.

If time allows, allow a child to step into your boots and pants and to put on your coat. Do not allow them to put on your helmet, however. A firefighter's helmet is too heavy for small necks. Also, head lice are endemic in some public schools, and you don't want to be even remotely involved in spreading the contagion. Most children enjoy dressing up in a firefighter's uniform, though, and will line up for the opportunity. Let them know from the beginning that there'll only be time for a few children to get their turn. You may wish to specify who and how many. So as not to appear to play favorites, one option is to dress up a child who is celebrating a birthday or whose birthday is the closest to the presentation date.

Usually I give handout material to the teacher at the end of the presentation rather than before. Distributing handouts early may siphon off some attention from the program itself. Handouts are valuable, however, in that they later remind the children of the visit and make it easy for them to go over the information with their family members. Handouts also let the parents know that a firefighter was at the school that day. Coloring books, bookmarks, and activity sheets are popular with the youngest age groups.

Upper Elementary School Programs

Upper elementary school children are more of a challenge than those in the lower grades. Older grade-school kids are acutely aware of their peers and don't want to risk being teased for being excited about or participating in programs. On the other hand, these kids can handle more challenging questions, and you can give them more detailed assignments. They have a greater sense of personal responsibility, and you can appeal to them by placing them in charge of their fire safety.

Home Escape Plan

Assign these students to design a fire safety escape plan for their family, and encourage them to teach their siblings and parents how it works. This gives them an opportunity to be viewed as knowledgeable and having some measure of authority.

Place the students in charge of teaching and reinforcing fire safety behavior. This will make them experts in the behavior you wish to reinforce, and they'll be motivated to learn the information thoroughly. Through their efforts, fire safety education will reach numerous people, including parents, who are often too busy to attend lectures on the subject.

Either have the students turn in the plans to you, or give the teacher a list of elements to check for, such as two exit routes out of each room, smoke detectors, and a designated family meeting place.

Instruct the students in the defensive techniques of crawl low under smoke and stop, drop, and roll, and have them practice these as well.

Instruct them also on cooking safety and how to treat a burn.

If you haven't met with this group for a few years, consider demonstrating your fire gear. Explain the function of each piece of equipment as you put it on. If you feel the group can handle it, have another firefighter enter the room in complete gear. The students will recognize him as a firefighter, but they may be a bit startled anyway.

Cooking Safety Class

Most junior highs and high schools offer cooking classes. Many young cooks learn the basics of food preparation there, so why shouldn't they learn the basics of cooking safety there as well? Cooking-related fires are among the main causes of fires in the United States today. According to the United States Fire Administration, cooking is the number one cause of residential fires in almost every state and locality. Nearly one-third of all residential fires originate in the kitchen. Since students at these levels are still in their habit-forming years, you should give them good, useful information that they can use toward following safe practices.

Planning

Call or visit the public and private junior-high and high schools in your area. Explain the importance of your program. Offer to present it at the teacher's convenience. The presentation takes about fifty minutes. Include all of the appropriate schools in your area.

Presentation

Bring a slide presentation, a skillet with a lid, a large mitt-type pot holder, a small amount of lighter fluid or similar flammable liquid to burn in the skillet, and the smoke detector demonstration unit as described in Chapter 14.

Introduce yourself, then explain that you're there to prevent the fires that you're normally called to extinguish. Impress upon the students the consequences of a home fire: that someone can get hurt or killed; the months needed to restore the damage; the displacement from home; the loss of time and money; the emotional cost.

Draw a fire triangle on the board. The term tetrahedron may go over their heads, so keep it simple. Tell them that, chemically, three factors must be present for a fire to occur. Ask whether anyone can name them. Someone will probably give you the correct answers. Give examples of how each ingredient contributes to a fire. After labeling the three sides of the triangle, explain that removing any one of the three factors will cause the fire to go out. Give some examples.

It's always useful to go over the subjects of carbon monoxide and the primary causes of fire in the home. Bear in mind that many students at this age are already smoking. Even those who aren't may have family members who smoke.

Career Day

Career day typically consists of a series of scheduled presentations, each lasting about a half-hour, by professionals in various fields. From my experience, the presentations can range from being downright boring to positively inspirational.

Nowadays, career day presentations are given at all grade levels, beginning at the preschool levels and continuing right on up through college. The common means of presentation include the captive audience format, in which a group sits as your audience for the length of what you have to say, and the booth, where people stroll by and talk to you as they please.

Preschools

The preschool version of career day is commonly called Community Helpers Day. A firefighter, police officer, and representatives of several other service providers each talk to the kids briefly about their careers.

For your presentation, you should begin by describing your job from the perspective of being a firefighter and, if applicable, an emergency medical technician. A suggested outline for the firefighting aspect of your discourse can be found later in this chapter.

In terms of the EMT material, let the children know that, although an ambulance may seem scary, it is really there to help people. Demonstrate some of the basic procedures for them, such as how to use a blood pressure cuff. Make the device seem familiar and nonthreatening. Tell them that the blood pressure cuff and stethoscope are designed to give a sick person's arm a hug. The medical technician listens to the hug with the stethoscope. Try the cuff on a willing volunteer.

Next, demonstrate the air splint. Have a volunteer step forward. Pretend that the volunteer has just had a bicycle accident, has hurt his ankle, and needs to be taken to the hospital so a doctor can take an X-ray. Explain how it's your job to apply the air splint to the foot to hold it straight so that it doesn't become more seriously injured. Show them how the air splint blows up around the foot like a big balloon.

Junior High and High School

Students at these grade levels are well on their way toward becoming functioning members of society. Some of them choose their career paths very early on. To help capture the interest of those who may be interested in a career in the fire service, outline a typical firefighter's workday. Explain the schedule of a professional firefighter. The notion of twenty-four-hour shifts followed by several days off is usually an instant hit with students. Take them through the daily work routine, including the checks of personal protective equipment; equipment maintenance; cleaning the station; EMT training; fire training; lunch; dinner; station tours; and free time to study, read, work out, or sleep, workload permitting. Point out that emergency runs take precedence over all other activities. Give them an idea as to how many emergency calls you might answer in a typical shift. Give a few examples of the types of calls to which you respond.

This last point is an important one, since many students don't realize that the modern firefighter is trained to perform a multifaceted repertoire of operations, ranging from EMS to fire investigation to rope rescue to haz mat responses, to name a few.

These kids'll be interested in money, so mention that, although salaries and promotion policies vary with the community and area of the country, most full-time firefighters make about $40,000 with vacation, sick, and personal time off during the year. In terms of personal requirements to make the grade, tell them that they have to be smart and physically fit. They must finish school, maintain good grades, exhibit good behavior, be in good physical condition, and avoid drugs. Test-taking skills are also important, so encourage them to begin acquiring them now. Recommend that they take advantage of all opportunities to take formal, standardized tests.

High school students will want additional information, such as where they can go to get training, how long the training will take, the entrance requirements, and so on, so be prepared to answer these questions with some firm information specific to your area.

Some of the questions that they may ask you include the following:

- How would you describe your work?
- What are the advantages and disadvantages of being a firefighter?
- Is there much pressure or stress associated with your job?
- What is the average starting salary and fringe benefits?
- How does the future look for your occupation?
- What special skills or interests do you need for your job?
- What in life or school prepared you for your job?
- What school subjects do you use in your line of work, and how do you use them?
- What personal characteristics should a person possess to be a good firefighter?
- What are the physical requirements to be a firefighter?
- Are there any hobbies or forms of volunteer work that I could engage in now that might help me do a better job as a firefighter?

- Are entrance exams or licenses required for the job?
- How does the work schedule affect family life?
- Why is this job important to you?
- What satisfaction do you get from your work?
- Are there opportunities for both genders within the fire service?

Short tapes and slides are useful when you have the time to use them. One way to reach students at this age is to show a video, set to music, that depicts a range of firefighting activities. The television studio of a local college or your local cable company might help you produce the video if your department doesn't have the resources for it. Your local cable company can also put the video on a continuous loop for you if you desire. This can be helpful when you're working a booth at a community event.

My department presents the pamphlet *The Firefighter NOW* to all career-day students of junior high school age and higher. A generic text for this brochure appears in Chapter 16. Use as much of the text and information in the pamphlet as you like.

Arson

When instructing young people on arson, the first task is to define it. Point out that some juvenile arson can be prevented by keeping fuels away from would-be arsonists. Since arsonists don't want to be seen setting a fire, other ways to forestall this sort of delinquency include having a clean yard and installing motion-sensor lighting. Owning a dog is also a useful deterrent.

As with other subjects of fire safety, some of the useful related topics include having a home escape plan, calling 911 in an emergency, smoke detectors, fire extinguishers, and survival techniques such as crawl low under smoke and stop, drop, and roll.

Fire Academy for Kids

At one point or another, most children dream of becoming a firefighter when they grow up. Who knows what sparks their interest? Maybe it's the sight of the truck, the flashing lights, or a brave firefighter battling flames to save a neighbor's home. Most children outgrow this fantasy, but some never do, and it is from among these that firefighters are born.

To help children better understand the life of a firefighter, members of my department teamed up with Schoolcraft College to present the Fire Academy for Kids as one of the electives of the college's Kids on Campus summer program. The program is open to children between the ages of eight and thirteen. It consists of two-hour sessions that run for four consecutive days.

The first day of class starts off with the presentation of some popular Hollywood movies that depict fires. After viewing the movies, we explain how the fire scenes were made. We then show these portions again so that the students can see the seams in the Hollywood illusion. Then we show a video of a real fire immediately after the movie scenes, thus punctuating the differences between a real-life fire and the Hollywood versions.

For kids, the high point of the fire academy is when they get to handle the hoselines.

After this introduction to the fireground, we outline for the students a typical day at the fire station. We then show them a complete set of personal protective equipment, including SCBA. A firefighter explains each piece as he puts it on. Later, the students are given a chance to handle the equipment.

On the second day of the program, we present a number of hazards commonly found in the home. This segment generally prompts a discussion about the home escape plan and the importance of devising one. We'll draw a sample plan on the blackboard, then assign the students to create one for their own home.

On the third day, the topics for the session are fire safety away from home and first aid. We tour the campus and learn about fire exits, fire extinguishers, smoke detectors, heat detectors, fire-pull stations, and sprinkler systems. All of these we describe in more detail when we return to class. The students learn how they can use this information to help them exit a building in case of an alarm. For the balance of the session, the students learn how to recognize and properly treat common medical emergencies.

The fourth day is the one that the children anticipate all week. This is the day they meet at the fire station, where we give them a tour and a demonstration of the apparatus. The students get a chance to walk through our fire training tower and to observe a training rescue drill. They then take over for a hydrant hookup and a hoseline stretch. The highlight of the day is when they handle the hoselines. All of them get a chance to handle the nozzle, and they have a tough time keeping dry.

Many departments have similar programs, and some are much more extensive than this one. If your department lacks such a program, it isn't difficult to implement, nor does it require much time or effort. You may choose to hook up with a college as we did, or you can initiate a program on your own. Either way, if you're not running one already, give it some thought.

Baby-Sitter Classes

An excellent but often overlooked way to spread the message of fire safety is to develop a baby-sitter class. If your department doesn't offer this training, you may be able to participate in an ongoing community program. The American Red Cross offers such programs in its community training centers, for example, and is usually cooperative in allowing the fire department to join in. A fire department can also team up with civic organizations, the police department, adult education programs, middle schools, the YMCA, and the local library to participate in existing health and fire safety programs. Many of these classes are accredited or certified by recognized agencies.

Contact these organizations and ask whether they offer baby-sitter awareness classes. If they don't, ask whether they can recommend an agency that does.

When you locate an organization that offers such classes, ask to speak to the individual in charge of the program and offer to teach the fire safety or first aid portions of the baby-sitter class. Other organizations are usually grateful for the services of qualified fire and EMS professionals.

When working out schedules, keep in mind that someone— maybe the person you're speaking to—has done a lot of research to get the program going. Therefore, be as considerate and accommodating as possible.

Working through an existing program eliminates the need for your department to get involved in a lot of start-up tasks, such as getting the program certified, finding competent instructors, locating a venue to hold the classes, advertising chores, and registration responsibilities.

Many baby-sitter programs run approximately eight hours. Make a record of how long your program takes and the material you cover. Such information should come in handy at a later date should you decide to run your own program.

I generally plan forty-five-minute sessions, which can be expanded to an hour by adding overhead transparencies, slides, or a demonstration.

The following sessions—plus anecdotes, question-and-answer periods, and audience interaction—should run about forty-five minutes each, for a total of an hour and a half of program time, which you can lengthen by incorporating visual aids.

Preparation

You should prepare for an audience that will most likely include boys and girls between the ages of twelve and fourteen. Practice in front of neighborhood children or family members until you become thoroughly familiar and comfortable with the material. You may also videotape your session in front of a make-believe audience. Keep the program lively. Audience interaction and humor can help keep the group attentive and sharp. If you are participating in another organization's program, be sure to show your outline to the class coordinator well before giving your presentation.

Content

The following are some of the key points that you should include in any program about baby-sitting.

- Be certain you feel physically and mentally fit to baby-sit.
- Bring along a baby-sitter's box containing emergency medical supplies, latex gloves, a flashlight, and some games to play with the children.
- Don't sleep on the job unless you have obtained prior permission from the parents.
- Don't listen to loud music, blast the TV, or use earphones that may prevent you from hearing the child cry out if he needs help.
- Before the parents leave, be sure to obtain from them the right phone numbers in case you have to reach them, their relatives, or their neighbors, plus any other pertinent information.
- Follow the parents' instructions and their house rules to the letter.
- Never leave a child alone in a bathtub or swimming pool. Always keep the children in sight.
- Keep the doors and windows locked, and do not open the door for strangers.
- Stay alert, and check on sleeping children frequently.
- Do not investigate strange noises outside. Call the police if you suspect trouble.
- If you receive an unwelcome phone call, be sure to hang up immediately. If you receive a second, similar call, telephone the parents or a neighbor. Don't let anyone know that you're a baby-sitter.
- When answering a call for the parents, tell the caller, "They're unable to come to the phone right now. May I take a message?"

Baby-Sitting and Medical Emergencies

• First aid is the immediate medical care that a layperson gives to an injured party before personnel arrive on the scene.

• If someone is sick or hurt, keep yourself calm, then try to calm the patient.

• If the injury or illness is serious, call 911 and the child's parents. Explain the type of emergency to the operator. Give your name and the address of the home at which you're baby-sitting.

• If the child has a nosebleed, have the child sit calmly. Pinch his nostrils gently, and tell him to breathe through his mouth.

• Cuts, scrapes, and minor abrasions can be painful, but they're rarely serious. Clean them with soap and water, then cover them with a dressing.

• For any other type of bleeding, apply direct pressure with clean dressing, and elevate the wound above the level of the heart. Don't let the child see the wound, if possible.

• Understand that choking is a serious medical condition and that more than 3,000 people in the United States die from it every year. If the child is coughing, encourage him to keep doing so. If the person isn't breathing or coughing, do the following. For an infant, use back blows and chest compressions. For a child or an adult, use abdominal thrusts. Always check for breathing.

When making your presentation, you should demonstrate the various blows, compressions, and thrusts. Explain how a firefighter would breathe for the patient if the person couldn't breathe for himself. Suggest that the baby-sitters each take a CPR course so they can practice how to perform these procedures correctly. I usually give a brief demonstration and tell them that the best way to learn the procedure correctly is to take a CPR course.

• For animal bites, wash the wound with soap and water. From a safe distance, try to identify the animal that bit the child. Call 911 and the child's parents.

• For human bites, wash the wound with soap and water. Call 911 and the child's parents.

• In case of a fall, do not move the victim unless he is in immediate danger. Call 911. Stay calm, and remain with the victim.

- In case of a bone fracture, move the child as little as possible. Call 911. Don't let the child see the wound, if possible. Splint above and below the break site if you must move the child.
- For sprains, elevate the limb and apply an ice pack on and off for the first twenty-four hours.
- A child may be poisoned by many substances around the house, including medications, household products, garden chemicals, auto products, and office supplies. If the child shows any indication of poisoning, don't delay. Call 911 immediately, then call the parents. Collect any pills or powders or whatever other evidence you have of the poisoning, and place it in a container. Keep the container sealed and give it to EMS personnel when they arrive. Call the Poison Control Center. Do not induce vomiting unless you are ordered to do so by medical personnel. Follow their directions.
- Suspect poisoning if you see open pill bottles, a half-eaten plant, or an open container, and if the child complains of a stomachache. If the poisoning doesn't appear to be life-threatening, call the local Poison Control Center. As many as 85 percent of all calls for poisoning are handled effectively at home.

Baby-Sitting and Burn Injuries

Review the common causes of fires, as well as basic fire safety procedures, including the need to crawl low under smoke and how to stop, drop, and roll. Since cooking safety has particular relevance to baby-sitting, review those principles as well. Be sure to incorporate the following points into your program:

- Never cook anything while holding a child.
- Don't allow any toys in the kitchen when you're going to cook.
- Test the temperature of all foods before you serve them to children.
- Don't let appliance cords hang down off the counters.

Baby-Sitting and the Home Escape Plan

- Ask the parents whether they have a home escape plan. If they do, ask whether you can practice it with the children when you sit for them. If the family doesn't have a plan, ask whether you and the children might develop one during your stay. If the parents say yes, set the children to work and keep them involved.

- When diagramming the escape plan, use crayons, markers, pens, or any other medium that interests the children. Have them indicate their rooms in color. Making this a fun and focused project will ultimately heighten their awareness of its purpose.
- Have the children assist you in testing the smoke detectors and in selecting the family meeting spot. If you discover that any of the smoke detectors aren't working, you should inform the parents before including them on the escape plan.

In formulating escape routes, the baby-sitter and children may need to incorporate an upper-story window as part of their plan. Let the sitters know that they can take out screens in an emergency by pushing on them. Demonstrate the hang-drop method of window egress.

Starting Your Own Baby-Sitter Program

If you decide to inaugurate your own baby-sitter program, begin by talking with the administration of your nearest American Red Cross or American Heart Association community training center. These organizations have printed materials ranging from handouts to textbooks that you can use as a guide toward developing your own.

You may want to include the following topics in your program:

- The care and handling of infants.
- Cooking safety.
- Food preparation.
- Kitchen safety.
- Symptoms of illness and caring for sick children.
- Law enforcement and the baby-sitter.
- Typical characteristics of children of different ages.
- Making up a baby-sitter's survival kit.
- Housekeeping.
- Home security.

Business contacts, coworkers, and spouses can help you prepare a baby-sitter's checklist. A local printing company helped with ours. We had con-

Baby-Sitter's Checklist

Parents' Names _____
Address _____
City _____
Telephone number _____

Children:
Name _____ Age _____
Name _____ Age _____
Name _____ Age _____

Where Can Parents Be Reached? _____
_____
_____
Time expected to return _____

Emergency Telephone Numbers:
Police
Fire   **911**
Ambulance

Poison control center _____
Neighbor _____
Relative _____

Discuss the Following With the Parents:
Meals and snacks _____
Bedtime or nap time _____
Medicine/allergies _____
Rules for TV and toys _____
Appliances and their operation _____
Burglar alarm _____
Smoke detectors _____
Fire extinguishers _____
Home fire escape plan _____
Possible safety hazards _____
First aid supplies _____
Pets _____

Notes: _____
_____
_____
_____
_____
_____

---

ducted an extinguisher demonstration at the company. The on-site coordinator appreciated our service and asked us to call her if we ever needed a favor in return. We requested help in livening up our baby-sitter's checklist. She arranged for free printing on heavy paper in flashy neon pink, with the Livonia Fire and Rescue symbol as a masthead. At our suggestion, the printing company's name appeared at the bottom of the checklist along with a thank you from us for all they donated.

Summer Safety Event

My department recently inaugurated a summer safety event similar to our Fire Prevention Week Open House. Called School's Out, Safety's In, the concept came from a similar program in Austin, Texas. We invite all of the school children in our community to attend the summertime safety party by distributing flyers in the schools, as well as through newsletters, newspapers, and public service announcements. The party is held on the first day after school lets out for the summer so that we can emphasize the safety tips for vacation activities in which the children will be engaging.

Our first such annual summertime safety party featured many of the

The bucket brigade is a sure-bet activity for the summer safety event.

activities featured in our fall Fire Prevention Week Open House. New activities and demonstrations were added to create a broader definition of safety, giving this event an identity all its own.

Many agencies and community members participated. A lifeguard from one of the city pools spoke about water safety and distributed pamphlets as reminders, as well as free passes to city pools. Local police officers staffed their DARE jeep and offered antidrug information to the attendees. The owner of a bicycle shop offered bike safety tips, and a rollerblade expert talked about safety for pavement skaters. A local library set up a table reminding the children of its numerous reading programs and special events scheduled for the summer, and local stores donated many giveaways, such as sports posters and movie rental certificates.

A lot of effort goes into an event of this magnitude, so it's best to start small. Be sure to provide a way for safety experts and guests to offer feedback on the event. Their comments can prove valuable when planning for the following year.

Chapter 4

Scouts

Webelos Readyman Badge

The Webelos-level Readyman merit badge of the Boy Scouts of America encompasses many emergency topics, such as emergency medical care; fire and burn safety; and car and bicycle safety. Suggested activities in the Webelos handbook include visiting a fire station and having a firefighter make a presentation, so don't be surprised if your department receives a few calls from group leaders.

A fire station is a great meeting place. The youngsters can assemble in a quiet area, such as a classroom or training area. The program should last no longer than two hours—about an hour for your presentation and about thirty to forty-five minutes for a tour of the station and the apparatus. Give the group leader an outline of your presentation before the scheduled meeting date so that he can brief the scouts on the topics that you'll cover.

When you introduce yourself, let the group know whether you were a member of the Scouts. The requirements for a Readyman badge—safety, medical care, treatment of burns and injuries, bleeding, and so on—relate directly to a firefighter's work, so this should help you establish an immediate rapport with your audience. Give examples of some of the common types of emergencies that would be phoned in to the dispatch center, and explain how they would probably be handled.

Call 911 for Emergencies

A foundation to give any safety-oriented group of youngsters is knowledge about the 911 system. The Webelos handbook gives scouts some information in this regard, such as what to tell the dispatcher. Have a few of the scouts practice calling in a make-believe emergency situation. Make sure they give complete information, stay calm, and don't hang up until the dispatcher tells them to do so.

Some additional pointers to tell the Webelos scouts are as follows:

• If you have a programmable phone, place the emergency number in its memory so that you only need to push a single button to make an emergency call.

• Use the 911 or other emergency number only in case of a fire, serious injury, medical problem, or other such emergency.

• Be prepared to tell the dispatcher what has happened, who is calling, and the location of the emergency—the who, what, and where of the incident.

• Don't move an accident victim unless it's absolutely necessary for the victim's safety, such as in a fire. Although movies and TV programs make it appear that every vehicle accident results in an explosion, explosions don't actually happen very often in real life.

The Hurry Scenarios

The Webelos handbook presents four hurry scenarios, designed to help these scouts recognize true medical emergencies and to take action expediently. If they aren't yet familiar with the term, these youngsters should know that, by definition, first aid is emergency medical treatment administered to injured or sick persons before professional medical care becomes available.

• Hurry Scenario One: The victim has stopped breathing. You can determine that the breathing has stopped by looking to see whether the victim's chest rises and falls; by listening for the air exchange in the victim's airway; or by feeling the victim's exhaled breath when you place your cheek beside his airway.

• Hurry Scenario Two: Blood is spurting from the victim's wound. This indicates arterial bleeding. Make a dressing with the cleanest material available, then put it over the wound. Wear latex gloves for protection whenever possible. Apply direct pressure to the wound. Elevate the area of the wound above the level of the heart. This will help to reduce the blood flow to the affected area.

• Hurry Scenario Three: The victim has been exposed to a poison. You must act quickly if you suspect poisoning. After calling for an EMS response, notify the victim's parents if they aren't at home. Try to find out what type of poison the victim might have taken, then try to collect some in its own container. Place the container in a bag so that it can be taken to the hospital with the victim. Having a sample will help the medical staff identify the poison more quickly.

• Hurry Scenario Four: The victim has had a heart attack. Some common signs of a heart attack include crushing chest pain; numbness or tingling in the left shoulder and arm that may radiate to the neck, back, or right side; weakness; pale coloring; shortness of breath; and a cool, clammy sweat. When any of these signs are present, call for an EMS response as quickly as possible.

Shock

• If the victim is weak, pale, cool, and clammy when you touch him, or if he is shivering, have him lie down. Raise his feet slightly and keep him warm. If the victim is unconscious, position him on his side. There is a section in the merit badge book that describes these procedures in detail.

Cuts and Scratches

• Thoroughly clean cuts and scratches with soap and water, then cover them with a sterile dressing. The dressing could be a simple adhesive bandage, or it could be dressing and tape.

• For all burns, immediately put cool water on the wounded area for at least ten minutes. While waiting for medical help to arrive, place a clean, dry bandage or cloth over the burn area after taking it out of the water.

Choking

• Performing the Heimlich maneuver or abdominal thrust quickly can save the life of someone who is choking. When the hands are in the proper position, they create a force needed to simulate a cough with the air in the lungs. Hopefully, this cough will be enough to drive the trapped food out of the person's airway. If you are the choking victim, know that you can perform this maneuver on yourself.

Safe Swimming

- Follow all safety rules.
- Never swim alone.
- Have an adult nearby.
- Make sure that life rings or other flotation devices are available.
- Avoid swimming when you're tired.
- Never swim with gum or food in your mouth.

Safe Bicycling

- Never ride a bicycle without a helmet. According to the National Highway Traffic Safety Administration, helmets reduce head injuries by 85 percent.
- Wear clothing that makes you visible to others on the road.
- Don't go bicycling at night. Most bikes aren't equipped for nighttime use.
- Always be alert for potholes, railroad tracks, gravel, and other hazards that could cause you to fall. Be especially careful when it's wet outside.
- Ride in single file on the right side of the road, going with the traffic.
- Check for traffic by looking left, right, left. More than 70 percent of car-bicycle accidents occur at driveways and other intersections.
- Bicycles are considered vehicles, so be sure to learn the rules of the road. Read the driver's handbook for your state.
- Check your bicycle for proper fit, snug wheel releases, and good braking.
- Never wear headphones while you're riding.

Vehicle Safety

- Always wear a seat belt while riding in a car.
- Never distract the driver with loud noises or arguing.
- You can assist the driver by looking out for emergency vehicles that have their lights and sirens on. If you see one, tell the driver immediately so that he can pull over to the side of the road until the emergency vehicle has passed.

Home Escape Plan

- It's important to create a home escape plan and to practice it with your family members. Be sure you have a designated meeting place outside the house.
- Tell your parents and siblings about defensive measures you can use in a fire, such as the importance of crawling low under smoke and how to stop, drop, and roll.
- Be the fire marshal in your house and make it your duty to ensure that all of the smoke detectors are operational at all times.
- Put together a home first aid kit in a can. Include basic items such as sterile pads, roller bandages, triangular bandages, latex gloves, and scissors, to name a few.

Tour of the Station

During the tour of the station, reinforce the safety lessons that you covered in your presentation. For example, point out that all emergency vehicles have seat belts, and even firefighters must buckle up while the apparatus is in motion. Identify the items in the first aid kit in the medical response units. Show the youngsters the equipment used to treat victims of external bleeding, shock, and heart attacks. Let them know about courses they can take in advanced first aid and CPR.

Cub Scouts

Every year, the leaders of the Cub Scouts in my area call to ask me to schedule a program designed to instruct children in the proper use of 911. The troop leader's guidebook suggests that the leaders plan ahead and arrange for a guest speaker from the fire department, so expect to receive an annual call yourself. Teaching this group what it needs to know about the emergency contact number won't take long, so plan to include some other fire safety information in the presentation.

We usually schedule the program for a mutually convenient date in the early fall at our fire headquarters, where we can host several groups at a time. Holding the meeting at the station has several other advantages as well, since the speaker is on his home turf and has access to all the resources there, including the ever popular fire trucks (which can be visited by the scouts) and firefighter protective equipment (of which you can give a demonstration).

A home escape plan homework assignment can be the criterion for earning the merit badge certification. The scouts' sketches should include all the components and potential situations listed in Chapter 3 for elementary school children. Describe some live-fire scenarios and demonstrate the hang-drop escape technique, as well as other methods of defense. Other topics of interest to the scouts are what to do until medical help arrives and the professional duties of a firefighter.

Fire Safety Merit Badge

This merit badge was called Firemanship until 1995, when it was given its current name. The handbook entitled "Fire Safety," a publication of the Boy Scouts of America, offers a very good introduction to fire prevention and safety. When preparing for your presentation, read this handbook to become acquainted with its contents. Doing so will help you explore ways of incorporating your background and experiences into the lessons to make them more lively.

Begin by discussing the requirements for earning the badge. As an extracurricular assignment, have each of the scouts devise a home escape plan and turn it in to you. Go through their plans and offer suggestions on how to improve them. You may wish to talk with the scout leader about this assignment well in advance of the program. That way, the kids can create their plans on their own, then turn them in to you at the beginning of the program, thus saving you from having to make a return trip for commentary.

Show some scenes from Hollywood movies about fires, then follow up with a videotape of a real fire in a real house, one that was perhaps taped with a remote recording device. Make them aware of the exaggerations and inaccuracies in the Hollywood depictions. This helps stress the point that real fires are quite different from the ones that young people are exposed to on television and in the movies.

A good topic to discuss with this audience is fire safety around the holidays. Point out to them that the Fourth of July is characterized by a plethora of cheap, easily obtainable fireworks being used by untrained amateurs. On Halloween, many of the hazards come from the candles in pumpkins and along walkways, especially given their proximity to autumn leaves. At Christmastime, there are even more potential sources of fire, given the number of open candles, highly combustible decorations, and electrical lights on trees. You might also mention that Christmas trees are often placed where they'll block an exit and that some people continue to put real candles on their trees.

Discuss the fire triangle and the fire tetrahedron. Demonstrate how a fire can be extinguished by removing any one of the three or four dependent components. Give specific examples in a real-world context, such as how you can put water on a burning chair to remove the heat, put a lid on a grease fire to remove the oxygen, or create a firebreak to stop a wildfire by depriving it of fuel.

The range of topics suitable for scouts working on their fire safety merit badge is an exhaustive one, covering such areas as carbon monoxide; burns; smoke detectors; fire extinguishers; crawling low under smoke; and stop, drop, and roll.

Merit Badge Day

My department receives numerous requests from various scout groups in the area to hold merit badge classes on fire safety and emergency preparedness. We've held many such classes at the school or church where the scouts normally meet, as well as at the fire station with on-duty crews. As happens at all fire stations, sometimes the on-duty crews have had to interrupt the program to respond to a call, leaving the scouts without an instructor.

Our department has several merit badge counselors, and it is they who came up with the idea of holding a Merit Badge Day—a meeting expressly designed to improve the delivery of the merit badge program. The objectives were to combine our efforts into a one-day program, given at our fire headquarters, where the scouts could also see the apparatus, have some fun, and earn their badge in fire safety or emergency preparedness.

Hours of preparation went into arranging for the event. Representatives from the Boy Scouts volunteered to handle all of the paperwork and contacts germane to their organization, allowing us to put all of our efforts into planning the program itself. At our first Merit Badge Day program, sixty-eight scouts attended.

Kicking off the presentation, our scenes from Hollywood movies juxtaposed against real-life fires gave the scouts an eye-opening view of what fire is really like. We followed this with a discussion of a home fire safety inspection. The scouts were assigned to complete, at home with one of their family members, one of our home fire inspection sheets, then return it to us. During the first break, the kids were given a tour of the fire engine and were permitted to squirt water from a charged hose. After the break, the class was split up into two groups: those who were seeking the merit badge in fire safety and those who opted for the emergency preparedness badge. After touring the rest of the station and other apparatus, the scouts were given addi-

The purpose of the Merit Badge Day program is to show scouts what fire, fire safety, and firefighting are really all about.

FIRE SAFETY MERIT BADGE ASSIGNMENT
WORK SHEET

Scout's name _____ Date _____

Troop _____

    1. Attach a newspaper or magazine article relevant to some of the common kinds and causes of fires in the United States.

    2. List some of the safety devices in your home and your family automobile, as well as some of the fire safety practices of which you are aware.

Home Safety Devices	Location
_____	_____
_____	_____
_____	_____
_____	_____

Vehicle Safety Devices	Location
_____	_____
_____	_____
_____	_____
_____	_____

Fire Safety Practices

_____
_____
_____
_____

    3. Devise a home fire escape plan, and practice it twice. The first time, follow the primary escape routes. The second time, use the secondary exits. On your plan, be sure to include all of the doors and windows, plus the primary exit routes, secondary exit routes, fire extinguishers, smoke detectors, and the disaster supplies kit.

    4. Have your parents assist you to make sure that you know how to do the following:

        A. Fuel a lawnmower safely.   Parent signature _____
        B. Start a barbecue.   Parent signature _____
        C. Test smoke detectors.   Parent signature _____
        D. Conduct a home fire escape drill. Parent signature _____
        E. Conduct a home safety survey.   Parent signature _____

tional training in their chosen subjects and were assigned homework that had to be turned in before their merit badges would be signed.

The additional homework for the emergency preparedness merit badge included preparing a first aid and emergency preparedness kit. Those seeking the fire safety merit badge were given various safety-related assignments, such as recognizing safety devices in an automobile and completing several activities under parental supervision. Examples of these latter activities included fueling a lawn mower, helping with a barbecue, and testing smoke detectors.

The media covered our first event. Those who helped organize the event held a postprogram meeting, which led to many ideas for improving the next event. Judging from the comments we received from the scouts and their parents regarding the homework assignments, the program was well received. Even though we didn't expect quite so many scouts to attend, we were still able to deliver a high-quality program that was entertaining as well as educational.

Additional Topics

Below are basic outlines for three areas of presentation of common interest to scouting groups: building safety, outdoor safety, and gasoline-powered equipment safety. Tailor these basic frameworks as you see fit to meet the needs of a particular merit badge program or other occasion.

Building Safety Survey

In your presentation, you should emphasize the following key points:

• Most fires and fire deaths occur in the home. Common sense would prevent most of these emergencies and the casualties that result. Too many fires are started by people who exhibit crazy, careless behavior. Once a fire gets started, people often worsen the situation by reacting inappropriately, taking risks, or delaying their call for help. Poor judgment and bad decisions only serve to increase the fatalities and injuries that result from emergency situations. Being armed with knowledge and having a plan of action beforehand can help to reduce these statistics.

• Various organizations publish information on identifying fire hazards in the home. Become familiar with the various hazards in the home environment on a room-by-room basis. Talk to your parents about these hazards and ways to reduce them. Devise a home escape plan and practice it with your family.

• Fire isn't the only threat to a home. If you smell gas, for example, check to see whether anyone is feeling ill, then call the gas company or the fire department.

• To be safer while in a building other than your home, take note of the exits when you enter. If you hear the fire alarm go off, take immediate action to evacuate and call 911 for help.

• The "Home Fire Survival" segment in the Boy Scouts' "Fire Safety" manual includes a fire prevention checklist that you can complete with occasional assistance from your parent or guardian.

Outdoor Fire Safety

Many important tips on this subject can be found in the "Fire Safety" manual under the section entitled "Outdoor and Camping Fire Safety." Some of the key points to make in an oral presentation are as follows:

• Outdoor and wildland fires burn thousands of acres and cause billions of dollars' worth of damage each year.

• Open flames, such as those of campfires and barbecues, must always be supervised by a responsible person.

• Always check to see whether campfires are allowed in the area where you're planning to build one.

• Build your campfire downwind of and well away from the tent. If the wind is blowing hard, don't build a campfire. Flying sparks from a campfire can cause a major disaster.

• Have plenty of tools handy for tending the fire, including full buckets of water, shovels, and rakes.

• Know how to use lighter fluid safely. Never spray it on open flame.

• Remember that the hot coals of a campfire can reignite days later. Before leaving the site, make sure that you douse the fire with water, stir the ashes, douse it again, and then cover it with dirt to kill the embers completely.

- Buy a tent made of flame-retardant fabric.
- Use flashlights for illumination inside the tent, never candles or matches.
- Know and follow the essentials of cooking safety as they pertain to cooking outdoors.

Gasoline-Powered Equipment Safety

- Store gasoline and other fuels in approved containers in the garage or shed, and make sure they're out of the reach of children.
- When filling the gas tank of a lawn mower or other machine, do so only when the engine is cool.
- Fill the tank in a well-ventilated area, preferably out of doors.
- Use only fuels that are recommended by the machine's manufacturer. Many people have been injured and killed as a result of putting the wrong fuel in a given piece of equipment.
- Remember that gasoline vapors are heavier than air and will go to low areas. If there is an ignition source present and the vapor density is high enough, an explosion or fire could result.

Chapter 5

Water Safety

Advance Planning

Water safety is an appropriate topic for various groups, including homeowners, resort owners, pool workers, students, and others. The resources that your department allocates to this issue will be determined by the amount of water sports activity in your community.

Planning ahead can help to decrease water-related emergencies and improve the chances of saving the victims of water-related accidents. Advance planning can include such activities as a swimming safety course for members of the community, a first aid course, and training in cardiopulmonary resuscitation (CPR). Preferably, the citizens in your area will take advantage of such programs as family activities.

Each year in the United States, more than 300 children under five years of age drown in residential swimming pools. More than 2,000 children younger than five are treated for water-related emergencies. Naturally, you should develop your presentation to meet the needs of your audience, but it should always cover three basic areas: supervision, barriers, and emergency procedures.

Supervision

- An adult must constantly watch children whenever they are in or near the water. Children are never invincible, no matter how well they can swim or how much they know about water safety rules. Most drownings in residential pools and spas occur when adult supervision has lapsed for just a moment. Almost fifty percent of the children who had fatal water-related accidents were reported to have been last seen inside the home just prior to the accident.

- Do not rely on inflatable devices to keep a child afloat in the water.
- No one should ever swim alone or in a body of water of unknown depth.
- Observe posted signs, rope barriers, and the instructions of lifeguards.
- If you don't know the depth of the water, don't dive. The water should be at least ten feet deep if you are going to dive into it.
- Don't swim in any body of water with which you're unfamiliar. Rapid currents and rocks can present unknown hazards in any body of water.
- Alcoholic beverages and swimming don't mix. Alcohol impairs judgment, balance, and coordination. More than half of all serious diving accidents occur when the diver has been drinking. If you haven't been drinking, but those who are with you have, abstain from swimming and diving. Convince your friends that it isn't a good idea to dive while under the influence of alcohol.
- Never swim during a thunderstorm.
- Don't swim after dark without adequate lighting.
- Never swim if you're tired or cold, if you've had too much sun, if you're on strong medication, or if you've been involved in a strenuous activity. Don't allow children to swim under these circumstances, either.
- Whenever you go on a boat, put on a personal flotation device that fits you correctly, even if you're a good swimmer.

Barriers

- Pool safety begins with proper barriers. Young children are naturally attracted to the water, and barriers can help deter some of them from wandering into danger. Still, you should never rely solely on barriers to keep children safe. There is no substitute for constant adult supervision.
- Make sure that the pool, spa, or hot tub has a safety cover or is enclosed by a fence or a wall that is at least four feet high.
- Many communities have ordinances with respect to pool fences. Find

out what they are in your community and be sure that your fencing conforms to the code.

• Doors leading to the pool area should be self-closing and self-latching.

• If your pool, spa, or hot tub has a cover, promptly drain any water that accumulates on the cover.

Emergency Procedures

• Keep a phone in the pool area so that you can call for medical help without delay. All family members should know the emergency phone number, and you should label it on the phone or post it somewhere nearby.

• Keep some equipment nearby that will help you make a water rescue. Examples include a long pole, ring buoys, and life jackets.

• If the incident involves a fall through the ice, do not go out onto the ice to try to rescue the victim. Attempt a rescue from shore, using a long pole, rope, or other such means. Call for help immediately. Tell the victim to remain calm and not thrash about. Even if you cannot reach the victim from shore, he may be able to squirm his way onto stronger ice by floating horizontally and kicking gently. Once on the ice, he should gently roll away from the hole. Coming toward shore, he should move slowly and keep his weight low, spread as widely as possible.

• Enroll in water rescue, first aid, and CPR classes.

Chapter 6

Industrial Safety

Industrial Fire Brigades

Both large and small industries in your jurisdiction may approach your department for help in starting an industrial fire brigade to respond to minor incidents in their facility. Some of them may be exploring an insurance carrier option to reduce insurance costs by establishing a brigade. No matter the reason, whenever a company approaches you, it is an opportunity to develop a successful partnership to work together to reduce fire losses and perhaps even gain support for some of your department's other programs.

Don't view these initiatives as threatening. The company often has very little information about the inner workings of your department and is merely trying to cut costs. If possible, arrange a meeting with the company managers to acquaint them with the services your department provides. Provide accurate information. If, for example, they ask you about your department's typical response time, keep in mind that they're not interested in judging your performance, but rather, in determining how much damage the company can expect to incur prior to the arrival of the first responders.

Actually, many fires in industrial environments are minor trash or machinery fires. If the company's brigade can extinguish such fires before the fire department arrives, hundreds of thousands of dollars in equipment may be saved, and the threat to human life will be diminished. Think of an industrial fire brigade as your first-arriving crew. Its members may extinguish the fire or perform other necessary operations before you arrive, such as removing people from the fire area or confining the flames.

Fire brigades tend to be comprised of either interested or assigned members trained in the basics of handling emergencies, such as evacuation, alarm systems, emergency communications, the use of fire extinguishers, and first aid. Some fire brigades are equipped with turnout gear, SCBA, and even fire apparatus.

Regardless of the size and complexity of their brigades, companies that initiate them offer your department an opportunity to participate in their safety training programs. Providing an instructor or two will help them get started with the brigade and other objectives. A mutually beneficial relationship can develop out of this. Your department will be able to establish relationships with individuals at the facility who can help you update your files on the hazardous materials and processes used at the site. Such information can be invaluable when planning for any response that you may make there in the future.

When planning the program, suggest to the contact person that the company allow forty to fifty minutes for the presentation and that at least the first half of the class be held in a conference area. Certain activities, such as fire extinguisher training, will involve going outside for a demonstration and hands-on practice exercises.

Hands-on learning is one of the most effective ways by which to acquire knowledge. As the visiting professional, you might be expected to help obtain extinguishers for use in the training program. One ten-pound dry-chemical extinguisher for every eight to ten students is usually sufficient. If necessary, suggest that the company ask its supplier to loan extinguishers for the day of the class. The extinguisher company's name and number should be on the annual inspection tag attached to the unit. If not, you can usually find the company in the yellow pages, listed under Fire Extinguishers.

Motivate the participants by pointing out that the information and skills that they acquire will help to promote safety at home as well as in the workplace, and that their newfound knowledge may even help them to save a life someday. Such objectives can draw in even those employees who may see themselves as only part-timers or who may be leaving the company in the near future.

Exits

All too often, people are only familiar with one or two exits in their workplace. This can be dangerous when the work environment consists of a maze of cubicles, confusing floor patterns, and the like. Getting your audience to see the need for clear paths leading to the exits and keeping them unobstructed will raise their level of concern for this simple but wholly vital aspect of fire safety.

Some other points to make are as follows:

• Check out all exit signs. The exit signs, which are usually illuminated, mark the way to enclosed stairs that lead all the way down and out of the building.
• Note the location of all of the exits that you might conceivably use during an emergency.
• Keep the exitways clear.
• Do not use elevators during a fire.
• If the facility is a hotel or a motel, consult the diagram of exits on every floor. As a manager or other employee of a hotel, you may be responsible for the safety of the guests.

Sprinkler Systems

• Sprinkler heads are activated by heat, and only those heads in close proximity to the fire will activate.
• If there is a fire, unexplained smoke, or an odor of smoke, immediately begin evacuation procedures, and summon the fire department by pulling an alarm or calling on the phone.
• Help any individuals who cannot exit on their own. Exiting coworkers can drag or carry to safety those fellow workers who are disabled or in shock.
• Confine the fire by closing doors on your way out of the building. This simple act can mean the difference between a room-and-contents fire and one that spreads throughout the entire building.

Industrial Extinguishers

• In many businesses, company policy or Occupational Safety and Health Administration (OSHA) regulations recommend or require fire extinguisher training.

• In an emergency, usually only a member of the fire brigade should use a fire extinguisher. Others can best serve the situation by exiting the building, calling 911 from an outside phone, and helping to account for the whereabouts of others.

• Know the locations of all extinguishers before a fire occurs. Check the extinguishers regularly, at least annually. If you find one that is defective or that needs to be charged, report it to the appropriate individual or department, such as a building maintenance person. Report any extinguisher that has discharged on its own, even if only for a short burst.

• Specific extinguishers are to be used for each class of fire. The type of fire for which each is to be used is designated on the extinguisher itself. Class A fires involve ordinary combustibles. Class B fires involve flammable liquids. Class C fires involve energized electrical equipment. Class D fires are those that involve combustible metals and are rare.

• There are two systems for marking extinguishers for use. The United States system consists of the applicable capital letter in a geometric shape. The International Picture Symbols have three panels, with a slash through any class for which the extinguisher cannot be used.

• Before using an extinguisher, do three things. First, ensure that everyone has left or is leaving the building. Second, be certain that someone is calling or has called 911. Finally, tell someone outside the building of your intentions to fight the fire, and have that person relay this information to arriving firefighters.

• Follow the acronym PASS whenever you use a fire extinguisher. The letters stand for pull, aim, squeeze, and sweep. First, pull the pin. Second, aim the nozzle at the base of the flames. Squeeze the handles together once within eight to ten feet of the fire. Finally, sweep the extinguishing agent from side to side at the base of the flames.

As the instructor, you should try to head off some of the problems generally associated with civilians using fire extinguishers in industrial emergencies. You can help do this by briefly describing fire behavior and by explaining that smoke and heat can spread quickly. Give them examples of how much fire you can control or extinguish with a handheld extinguisher, and make them aware that the canister will quickly run out of extinguishing agent. Emphasize that they remain aware of their escape routes from the fire area and that someone call the fire department without delay as soon as the emergency is recognized.

If practical, ask your fire prevention division to conduct the outside fire extinguisher training. For a fire burn pan, you can use a galvanized oil drain pan from a local auto parts store. Consider using a long-handled lighter, protective gloves, newspaper, and a touch of fuel to allow you to reignite the fire easily. Place the newspaper in the pan, fill the pan halfway with water, then add a small amount of diesel fuel. The diesel fuel will float on the water and be heated by the burning newspaper.

When selecting a location for the demonstration, consider the direction of the wind. Assemble the group with the fire extinguishers upwind. Move all vehicles downwind, far enough to keep any overspray from reaching them. Always keep an extinguisher close by in case a problem should arise.

Once a student has discharged the extinguisher for the first time, expedite your training of the rest of the group. The propellant seeps out quickly, and if you don't keep the pace up, you'll be left with a useless extinguisher and some unhappy participants.

Finally, don't force anyone to use the extinguisher. Suggest to those who don't want to take a turn that they at least view the demonstration and see how effective a portable extinguisher can be. If enough time remains afterward, offer the reluctant ones another chance. If they accept, stay close by to offer encouragement.

After the hands-on demonstration of industrial extinguishers, call your audience's attention to home extinguishers. Display several. Mention that an ABC extinguisher is recommended for homes, since it can be used for Class A, B, and C fires, all of which can occur in the home. Point out that home extinguishers are smaller than their industrial counterparts. Consequently, they operate for a shorter period of time—seven or eight seconds versus one minute for an industrial model.

By way of closing, give a short talk to your audience about smoke detectors.

Chapter 7

Medical Emergencies

First Aid

If you think that the citizens in your community might be interested in a first aid class, develop one. Not every program that you introduce has to continue forever. If the class isn't well attended, you can discontinue it or schedule it less frequently.

Department members who will teach the class should be certified through the nearest American Red Cross or other training facility. The American Red Cross is closely associated with first aid training and would therefore lend additional caliber to your program.

Allow enough time to plan and promote the class. Advertise it, and encourage advance registration so that you'll be able to gauge how many books, handouts, and other materials you'll need. Advance registration also allows the instructor to focus on teaching rather than enrollment on the first day of class.

Advertise in your local newspaper. Distribute flyers, posters, and other literature in public areas throughout your community. List the class on your cable company's bulletin board.

In the meantime, clean up your manikins, and prepare your other classroom materials as well. If you wish to save time, you can purchase handout materials from the Red Cross. To keep costs to a minimum, make your own handouts on a computer.

As you should always do whenever you intend to give a presentation, prepare and follow an outline, for it will help you stay on track. Keep the language in your classroom simple. Don't use jargon that the audience might not understand. Never talk down to your audience. Finally, prepare an evaluation form that class members can fill out anonymously at the end of the presentation. Their feedback will prove helpful toward fine-tuning future programs.

Emergency Action

If you find that a complete first aid course will consume too great a portion of your department's time or resources, you might want to substitute an abbreviated version. Also, the thought of committing themselves to attending an American Red Cross First Aid course or an Advanced First Aid course might seem intimidating to some citizens. An alternative might be to start a one-hour class that can serve as a primer to first aid. A descriptive title for this sort of short-form course could be "Emergency Action: What to Do Until Help Arrives." This sort of class can be held at an industrial site, the fire station, the local library, or some other community facility.

Use videotape segments, slides, or overhead transparencies as visual aids. Create handouts that have plenty of margin space for notes and comments. Explain to your audience that the objective of the class is to provide personal knowledge and that taking it won't lead to certification in first aid or CPR.

In your presentation, you should emphasize the following key points:

• The emergency medical service (EMS) system is comprised of community members who are usually the first to know when an emergency has occurred.
• Time is of the essence in a medical emergency.
• You may be the first and only witness to a medical emergency. Act quickly. Dial 911 to put the EMS system in motion. Give accurate details about the incident so that the dispatchers can send the proper equipment.
• Before approaching the victim, make sure you aren't placing yourself in danger. There may be hazards from traffic or electrical wires, for example.
• Stay calm. Your calm demeanor will help to soothe and reassure the

patient. Even if the injury scene is the most horrible situation that you've ever encountered, pretend that it isn't that bad. Try to shield the patient from any obvious deformities.

- Do not move the patient unless it is absolutely necessary to protect him from further injury.
- This may be all you need to do. However, if you would like to contribute more to your community and believe that you have the temperament and time, the emergency services always have a need for those who have trained eyes and ears and can report emergencies correctly.

When giving a talk on first aid, leave time for a question-and-answer period. Explain to the audience that, by definition, first aid is the immediate medical care given to a patient by a lay person before medical personnel arrive on the scene. For those seeking knowledge beyond the scope of the class, suggest that they take an advanced course or watch a video on first aid and emergency medical procedures. These videos may be available on loan from many video rental stores, the public library, or the local fire department.

CPR Classes

Conducting classes in cardiopulmonary resuscitation (CPR) can be a great, inexpensive way to reach out to your community. A CPR class covering treatment appropriate for an adult takes about three and a half hours. A complete CPR course, covering treatment for the infant, child, and adult, takes about eight hours. Such a class can be given on a single day, perhaps on the weekend, or it can be spread over two weeknights.

The American Red Cross and the American Heart Association are among the primary organizations offering instructor certification courses. To register for the Red Cross program, contact the nearest chapter. For the American Heart Association, make arrangements through its community training center (CTC) in your area. These centers are often affiliated with the local hospital's community health education department.

Have prospective attendees preregister for the program so that you'll know how many instructors and how much material you'll need. Schedule classes through your affiliated organization, and promptly return the requisite student forms after completion of the class.

Chapter 8

Evacuation Drills

Houses of Worship

Houses of worship sometimes request assistance in establishing an evacuation plan for their periods of peak occupancy. The plan should include both the actions to be taken in an emergency and provisions for training congregation members in these actions. The ideal approach to developing such plans is to walk through every house of worship in your district. This could prove time-consuming, but it would also provide the highest level of benefit.

In addition to training the clergy and staff of a given institution, you should also provide some basic instruction in the use of fire extinguishers. Since church- and temple-related arson is on the rise, discuss with the congregation members ways to reduce this crime, such as organizing a neighborhood watch program. Congregation members can drive through the parking lot whenever they're in the area and report any suspicious activity to local authorities. Ask neighbors who live near the house of worship to be more aware of strangers and vehicles in the area. To reduce other forms of crime, encourage the institution to engrave any valuables with an identifying number and to photograph or videotape all of its property. If the house of worship doesn't have a burglar alarm system, its management should consider installing one, and the presence of such a system should be announced by way of posted notices on the doors and windows. Timers and photoelectric devices are also useful. A motion detector in proximity to the shadowy regions around certain windows and doors can illuminate these areas if an intruder trespasses there at night.

The basic points to make when encouraging a house of worship to devise a basic evacuation plan are as follows:

- Create a map of the institution that clearly indicates two emergency exits and the designated meeting place.
- Post a copy of the map outside the designated meeting rooms in each area of the institution. When more than one meeting is planned, not all of them should be held in the same area of the building. Smaller numbers of members should meet in distal rooms so that evacuation and accountability can be facilitated in an emergency.
- Designate able congregation members to help nursery workers evacuate the infants and children during an emergency.
- Discuss and have a plan for the evacuation of the elderly, the deaf, the physically disabled, and others who may face compounded threats during an emergency. Those who are to be designated helpers should have some knowledge of drags, lifts, and other methods of physical evacuation of the disabled.
- Occasionally practice methods of alerting the members to evacuate. Sound a fire alarm, for example, or issue a practice evacuation call over the public address system.
- Remind the members to close interior doors behind them as they exit the building, since this will help to compartmentalize the fire.
- For the staff and maintenance personnel, observe good housekeeping rules. Keep combustibles out of the building, trim shrubs well below the level of the window, and keep tree branches higher than the windows.

Schools

Schools conduct fire exit drills to ensure that students will use the available exits to leave the building quickly when the fire alarm sounds.

Maintaining order and control is imperative. These objectives can be achieved only when the students are familiar with the procedures, which should be practiced until every student knows exactly what to do from any point in the building if the alarm goes off.

Some states have laws that hold a specific agency responsible for scheduling, conducting, and determining the number of fire drills. If no such law or code is in place, the school district may follow the recommendation of the International Fire Service Training Association (IFSTA) to hold at least two fire drills during the school year. IFSTA further recommends that one of the drills be held early in the school year to train the students and that the second be held after the holiday break to reinforce their drill behavior.

School administrators should abide by the following procedures for fire safety and evacuation drills in the school:

- Develop a fire escape plan that includes a primary and a secondary exit, as well as a predetermined outside meeting place for each class.
- For any fire drill, sound the actual fire alarm. Don't initiate a drill by sounding the bells normally used to dismiss class.
- Call the fire department whenever a nonscheduled alarm is sounded at the school.
- Inspect school exits daily to ensure that doors are unlocked, exits are unblocked, and all of the stairs and hallways are accessible by all.
- Assign some of the more mature students to act as monitors who'll assist in the evacuation.
- Teachers and other staff members should search the restrooms and other seldom-used areas, as well as those with limited access.
- Once outside, have the students stay there until they have been notified that it's safe to reenter the building.

If your department is in a community that has regularly scheduled school fire drills, look for ways to assist. Offer to come to the school for one or two of the drills. This lends an air of credibility and seriousness to the drill. Offer to talk to the students in the auditorium afterward to emphasize the importance of responding to the alarm promptly.

When speaking to an audience of older students, several examples can dramatically emphasize the need for school fire drills. The Our Lady of Angels School fire in Chicago and the Lakeview Elementary School fire in Collinwood, Ohio, are examples of how improper exit procedures can lead to numerous casualties. In the Lakeview Elementary fire, 175 people died. That fire occurred in 1908. Fifty years later, 95 people died in the Our Lady of Angels fire. This school was of ordinary construction, with open stairways that allowed heat and smoke to spread rapidly. It was equipped with neither a fire detection nor a fire suppression system. These examples show that fire safety must be engineered into the building itself, as well as practiced by those who will occupy it.

Health Care Institutions

Targeting a public fire safety education program toward health care institutions is a task that is often overlooked as we keep busy meeting the requests of civic groups, senior citizen groups, schools, and other organizations that regularly ask for such programs. Still, the special needs of health care institutions must be addressed.

As is the case with most businesses, health care workers generally do not put fire protection or fire suppression activities high on their list of daily duties. It is up to us to remind them periodically that fire doesn't only visit those who are prepared for it. Large hospitals often have in-service training programs pertaining to fires, but smaller clinics, nursing homes, and group homes may not.

Most hospitals require each shift to conduct fire drills quarterly. Scheduling your program to coincide with one of these drills will enable you to tie into the hospital's fire safety program and meet with the staff while the drill is fresh in their minds and patient safety is the topic of the day.

The acronym RACE (rescue, alert, confine, extinguish) is a key concept for fire safety programs in health care institutions. Ideally, the rescue and alert phases will occur simultaneously. If not, the individual must make a judgment call as to which function to perform first. A good rule of thumb is that if the staff member has to leave the fire area to get help but someone in that area is threatened, then rescue must occur first. Life safety is the primary concern of health care workers, and they must act accordingly. Calling out to your coworkers for help while performing a rescue is usually a good idea.

The principle of rescue applies just as much to guiding an able-bodied person toward safety as it does to helping the infirmed out of an involved room. Assistance can range from opening a door or a window to carrying an unconscious person down a flight of stairs.

In teaching rescue, demonstrate proper lifting techniques. Practice various ways of lifting, dragging, and carrying patients. A practical rescue over a long distance involves controlling the victim in a headfirst drag. This can be accomplished by placing the victim on a blanket or by grabbing his clothes firmly. If you drag the victim feet-first, you'll risk having his arms move up over his head, catching on objects such as doorways and furnishings. Also, this would make it difficult to negotiate stairs. If the beds have wheels, it would probably be easier to move the patients in their beds.

Alerting others to the emergency can mean anything from calling 911 to pulling an alarm to calling out a window. Some larger institutions use the public address system to spread the word within the building. So as not to induce panic among the patients or residents, sometimes these messages are in code, giving the staff enough time to initiate organized evacuation procedures. Make sure your audience members know the meaning of the message, which may be something like "Paging Doctor Smoke, third floor, west." Even if no bell or other conventional alarm has been sounded, the staff should begin the evacuation immediately on hearing the code word.

If there have been instances within your district in which fires have resulted in the complete destruction of structures because of a delay in notifying the fire department, you should cite them for your audience, with details.

In learning to confine the fire, health care employees should be familiar with the concept of defending in place; i.e., knowing the fire zones within the building and having the occupant wait out the fire if escape is likely to prove unsuccessful. Workers in the facility should know which items they can use to help them in this strategy. Some to mention are the locations of the fire zones, telephones, and outside escape ladders. They should also know which windows lead onto lower roofs, as well as how to pull the alarms and use the public address system. In addition, they should know the types of fire extinguishers in the facility, their location, and how to obtain sheets, towels, linens, and water quickly.

The final function of the RACE acronym, extinguish, is an option only for those who are competent in using a fire extinguisher and who know that the evacuation is complete or nearly so. No individual who is uncomfortable using a portable fire extinguisher should be forced to use one. Fire extinguishers should only be used against small fires and only if no one, including the user, is placed in jeopardy. Just as you wouldn't try to stop major arterial bleeding with a finger bandage, you shouldn't try to put out a sizable fire with a portable extinguisher. Ensure that your audience knows how to select the proper class of extinguisher. Pass around an extinguisher from the health care facility so that the members of your audience can become familiar with it.

Emphasize that extinguishment is an option only after the rescue, alert, and confinement functions have been fulfilled. Also, someone outside the building must know that someone is inside fighting the fire so that he in turn can notify the incoming fire companies. The person inside should always keep an exit to his back when using an extinguisher in case the unit fails to operate or cannot control the fire effectively.

The PASS acronym is another that your audience should remember: pull, aim, squeeze, sweep. Mention that they should expect to feel a slight recoil when they activate the extinguisher. The unit will make a noise as the agent rushes out, especially the carbon dioxide type.

Give your audience an opportunity for hands-on operation of the extinguisher, if possible. The base of a fifty-five-gallon drum, cut one to two feet high, makes a nice vessel to hold whatever fuel you use. Paper may be a good fuel, but a bit of flammable liquid will help the fire restart faster after you've extinguished it. Check with your fire marshal, training officer, and dispatcher before using fire as a teaching tool. Always have a backup plan in mind in case of emergency.

Staff turnover at health care facilities is often high. Consider making a videotape of your presentation and leaving it with the management so that it can be incorporated into the employee orientation program.

College Dorms

One of the biggest problems in teaching fire safety to the residents of college dormitories is that your audience essentially consists of transients. Many of them are experiencing life away from home for the first time and typically don't consider emergency preparedness to be any kind of priority. Still, college dorm fires have taken numerous lives in recent years. Unfortunately, such incidents tend to claim several people per fire. This makes safety programs in college dorms an important part of your fire department's action plan.

Living in a college dorm, a student's safety is constantly under the threat of the actions of others living there. Contact the colleges in your community, and schedule a meeting with the appropriate person. The head of security is often a good place to start.

When addressing an audience of dormitory residents, the essential points to cover are as follows:

- Have a home escape plan, and practice the procedures at least twice a year.
- Be familiar with the fire alarm system.
- Know the locations of the fire extinguishers.
- Maintain fire protection equipment. Determine the type of detectors present in the dorm. Find out how they work and whether they need to be maintained. Check with the college about the testing and maintenance program. Who should test the detectors in a dormitory? Smoke detectors should be installed in each bedroom and in common living areas. Detectors placed in halls may not always activate if there is smoke in a bedroom or other chamber. Check the extinguishers to ensure that they're in operating order, haven't been discharged, and have had the regularly scheduled maintenance test.
- Report any missing or malfunctioning equipment to the appropriate person.
- If you're allowed to cook in your dorm, practice safe cooking procedures.
- If you smoke, do so safely and responsibly. Do not smoke in bed.
- Never leave candles unattended. Blow them out before leaving a room, and always use candle holders. Use common sense when it comes to open flames of any kind.

- Know that electrical safety is paramount in college dorms. Typically, there are too few outlets for so many computers, stereos, blow dryers, and other appliances. Don't overload electrical outlets. Only plug one appliance into an outlet at a time. If there simply aren't enough outlets in your room, use rated and approved power strips rather than the octopus outlets so often found clumped and hanging from the walls.
- Don't place curtains or clothing over lamps or lampshades. Some bulbs can reach temperatures of more than 1,000°F.
- At parties, make sure that someone stays sober and keeps watch over the rest.

When you're finished giving your presentation, schedule another meeting so that you can address those students who were unable to attend the first time. As you become more comfortable with this type of audience, expand your presentations to include additional material, such as how to deal with medical emergencies.

Hotels and Motels

According to the NFPA, the percentage of hotels and motels that have sprinkler systems nearly tripled between the years 1980 and 1993. Still, the benefits of these appliances are often nullified because the staff may not know what to do in an emergency. Additionally, few guests bother to note the basic safety procedures during their stay. Many of them will use elevators during a fire, and those leaving their rooms may not close the door behind them or bring their room key.

The Las Vegas Fire Department has addressed this problem by instituting a hotel life safety program. The plan includes assigning hotel and motel employees to one of five emergency teams. Each team has a different assignment during an emergency. The hotel executive team coordinates activities between the hotel staff and fire department personnel, maintaining as much as possible the continuity of business within the hotel. The fire control team pinpoints the location of the problem and briefs the responding fire crews through the executive team. The emergency response team goes to the scene of the alarm, verifies the alarm, and takes the appropriate actions according to the conditions they find and the level of training of the team members. The evacuation team helps with the orderly evacuation of the guests and leads them to a location away from the front door. Finally, the damage control team assists fire personnel with salvage operations during firefighting operations.

A good place to start when approaching professionals in this industry is to talk about each of these teams, then let the management of the hotel decide whether it wants to implement such a program. If the managers agree to it, they should assign responsibilities to each employee and position. Other assignments are contingent on the individual property. Work with the staff to help them establish the program. It may be necessary to hold additional meetings to train them.

If the management doesn't want to implement this type of program, you may offer other types of training, covering topics such as how to use a fire extinguisher, what to do until help arrives, and how to deal with medical emergencies.

Chapter 9

Adult Education

Senior Citizens

Presenting fire safety education to the elderly isn't without its challenges. Their aging bodies prevent them from doing many things they could once do with ease. Even climbing out a window or learning how to stop, drop, and roll may be beyond the ability of some.

To schedule fire safety programs for seniors, you can find the contacts you need at municipal senior citizen centers, senior housing unit associations, your local chapter of the American Association of Retired Persons (AARP), and religious institutions.

When approaching an audience of the elderly, you should emphasize that senior citizens are at high risk for injury and death in fires. Along with children, Americans sixty-five years of age and older have the highest death rates from fires occurring in the home. Among the reasons are that seniors often take medications, which may make them sleepy; they often live alone; and they aren't able to move as quickly as younger people.

Additionally, a large number of men and women over the age of seventy-four in this country die in fires associated with cooking. Often these fires are caused by combustible items lying too close to open flame. In many cases, the cook's clothing catches fire. The United States Fire Administration reports that cooking has been the leading cause of residential fires during most of the years since the inception of the National Fire Incident Reporting System (NFIRS). Obviously, many cooking fires and related injuries go unreported each year. In a large share of them, senior citizens are involved.

Many senior citizens remember the days when smoke detectors first became widely available, back in the 1970s. They cost about fifty dollars, which was quite a bit of money back then. Many don't realize that the cost of these units has since dropped considerably. Be sure to point out to them that smoke detectors now go for as little as five dollars. You should also demonstrate a smoke detector for this audience, showing them how a modern unit responds to very little smoke.

Some departments have smoke detector programs for the elderly, in which they supply and install the units free of charge. If your department has such a program, it should be widely publicized. If not, you should consider the benefits of implementing one.

Other appropriate topics for people in this age group include cigarette safety; home escape plans; space heaters and other heating equipment; electrical appliances and cords; and fire extinguishers. As a lead-in to your presentation on fire extinguishers, consider telling a story or showing a video about a fire that was successfully extinguished while it was still small. Discuss with your audience some steps that they could take right now to ensure that they don't become casualties of a fire themselves.

Families After the Fire

A family whose house has just suffered major fire damage often stands by not knowing what to do when every belonging they owned is now burned beyond recognition. Firefighters can help these families. Your department may hand out a fact sheet that explains what the family should do next, but some of the sheets I've seen don't answer the majority of the questions that homeowners typically have. The booklet "After the Fire, Returning to Normal," available without charge from the United States Fire Administration, outlines actions that a family with insurance should take. It also tells them what items are typically covered by insurance, how to evaluate the property to inventory the loss, and what to do if you don't have insurance. It also contains a handy checklist of agencies to contact when documents and records must be replaced. Still another helpful feature is the section on salvage hints, part of which explains how to get the smell of smoke out of furnishings and clothing. Your department could personalize the USFA booklet rather than create one of its own. Be sure to include the name of the original publication and a USFA reference.

In our community, for those residents who are burned out of their homes by destructive fires, our department presents coupons for free lodging donated by local hotels.

A fire is devastating to a family. A little compassion can go a long way. Remember that you, as a professional firefighter, are a public servant. Offer those people a few minutes of your time, give them some meaningful information, and help them if they need additional assistance.

Life Facts Information Card

How many of the cards in your wallet can help save your life in an emergency? The Livonia Fire and Rescue's Life Facts Information Card is one that can.

The concept of such a card grew out of a number of ideas, the most common of them being the school system's Emergency Contact Cards, which parents are asked to complete on a regular basis. In our ongoing commitment to keep the community safe and informed, Livonia firefighters took this concept one step further.

The Life Facts Information Card is designed to provide concise personal, medical, and emergency contact information. Small enough to fit into a pocket or a purse, it can give responders the information they need to administer emergency medical care or obtain the necessary authorization to do so. Since the cards are printed on cardboard and are wallet-size, their cost is minimal, thereby making it possible for you to give them away free to area residents, civic groups, schools, and anyone who may visit a fire station. All of the information asked for on the card is optional and can easily be changed.

Emergency Contact Information

Name _____ Relationship _____
Home no. _____ Work no. _____
Cell phone _____ Pager _____

Name _____ Relationship _____
Home no. _____ Work no. _____
Cell phone _____ Pager _____

Name _____
Home no. _____
Cell phone _____

Life Facts Information Card

Name _____
Address _____
City _____ State _____ Zip _____
Date of birth _____ S.S. no. _____
Family doctor _____ Doctor's phone _____
Health insurance co. _____ Policy no. _____
Blood type _____ Medications _____
Medical conditions _____
Allergies _____ Surgical history _____

For elderly citizens, who often keep important medical information in various locations about their homes, the convenient wallet-size card can hold all of the information that responding firefighters need to administer proper emergency care and complete an incident report. For a child who is injured while away from home, it may mean that the responders can contact his parents more readily if the information is on his person or in his backpack. Conversely, parents who are away can fill out the card and leave it with grandparents, baby-sitters, or neighbors.

Adding this kind of information card to the services you provide can make the citizens of your community feel more secure when they or their families need emergency care.

Chapter 10

Seasonal Safety Programs

Calendar of Events

You may wish to expand the scope of your fire-related safety messages into other realms. The following calendar suggests themes for community programs throughout the year. Establishing partnerships with other organizations may help you spread your safety messages.

January
- School Nurse Day.
- National Eye Health Care Month.

February
- National Child Passenger Safety Week.
- National Children's Dental Health Month.
- National Burn Awareness Week.
- American Heart Month.

March
- National Poison Prevention Week.

- National Nutrition Month.
- Red Cross Month.
- National PTA Alcohol and Other Drugs Awareness Week.

April
- Keep America Beautiful Month.
- National Child Abuse Prevention Month.
- National Infant Immunization Week.
- Reading Is Fun Week.
- World Health Day.
- National Youth Sports Injury Prevention Month.

May
- National Physical Fitness and Sports Month.
- National Emergency Medical Services Week.
- National Transportation Safety Week.
- Buckle Up America Week.
- National Bike Month.
- Safe Kids Week.
- Clean Air Week.

June
- National Safety Week.

September
- National Farm Safety and Health Week.
- National Head Lice Prevention Month.
- Children's Eye Health and Safety Month.
- National Cholesterol Education Month.

October
- National Fire Prevention Week.
- National Child Health Month.
- Family Health Month.
- National School Bus Safety Week.

November
- National PTA Child Safety and Protection Month.
- Great American Smokeout.

December
- National Drunk and Drugged Driving Prevention Month.
- Safe Toys Month.
- World AIDS Day.
- National HIV and AIDS Awareness Day.

Spring and Summer Safety

Everyone likes to go out when the weather turns warmer. Following are some of the topics worthy of mention when teaching safety for spring and summer activities.

Motor Trips

- Every year, approximately 1,400 children are killed and 300,000 are injured as a result of being a passenger in a motor vehicle.
- The correct use of safety seats reduces the risk of death by 71 percent, hospitalizations by 67 percent, and minor injuries by 50 percent, according to the National Highway Traffic Safety Administration.
- While traveling in a car, always restrain a child in the appropriate child safety seat or with seat belts. On average, more than half of the children under the age of four killed in crashes each year are unrestrained, according to the U.S. Department of Transportation.
- Child safety seats fall into three general categories. According to the U.S. Department of Transportation, infant-only seats are designed for infants from birth to approximately twenty pounds and twenty-six inches in length and under one year of age. Convertible seats, the second type, are used either in a rear-facing configuration for infants or in a forward-facing configuration for children of more than one year of age, weighing at least twenty pounds and up to forty pounds. Finally, booster seats are used for older children who might have outgrown standard child safety seats and

are transitioning to adult safety belts. Booster seats are best for children who weigh from forty to sixty pounds.
- Never use a child safety seat that has been involved in a crash.
- Always secure the child correctly, per the manufacturer's specifications, in a child safety seat.
- Never use pillows or cushions to boost a child. This may cause him to slide under and out of the seat belt.
- Never put luggage or other hard items on the shelf of a car's rear window. Any items placed there could fly forward and injure passengers if the driver has to make a sudden stop.

Backyard Barbecues

- Choose a safe spot for the barbecue grill, if it is portable. Take into account the wind direction and the paths that children use.
- Keep the barbecue away from combustibles such as leaf piles, overhangs, and houses.
- Tell children not to go in the area of the barbecue grill whenever the grill is hot, whether or not food is being cooked.
- Never leave the grill unattended, and be ready to extinguish accidental fires by having a garden hose or a bucket of water nearby.
- If you use charcoal, use only charcoal lighter fluid to ignite it.

In-Line Skating

- Fifteen children have died from in-line skating accidents since 1992. According to the U.S. Consumer Product Safety Commission, approximately 65,000 children ages fourteen and under were treated for skating-related injuries in 1995 alone.
- In-line skating involves the mastery of five basic skills: balancing, proper form, stroking, braking, and falling.
- The best way to learn how to skate is to take lessons through skate shops, sports stores, or fitness centers.
- Always wear full protective gear; namely, a bicycle helmet, knee pads, elbow pads, and wrist guards.
- Stay out of the street.

- Skate in areas that are free of sand, gravel, and dirt.
- Avoid hills.
- Don't skate at night.
- Never wear anything that affects your hearing, such as headphones.

Fireworks

It's hard to believe that fireworks are sold to people who are untrained in their use. The injury statistics related to over-the-counter fireworks are staggering. Fireworks are entertaining, but only when nothing goes wrong. For this reason, I recommend that citizens attend public fireworks events and not attempt having home displays.

Fireworks, designed to explode and throw a shower of sparks in different directions, reach surface temperatures as high as 1,200°F. On a typical Fourth of July, more fires in the United States will be caused by fireworks than by all other causes combined.

In 1995, fireworks were involved in 27,400 fires in the United States. According to the NFPA, these fires killed one civilian, injured another ninety-three, and caused $32,500,000 in direct property damage. Most of the reported injuries were burns, contusions, abrasions, and lacerations. When you take into account injuries that aren't reported to fire departments, the numbers are significantly higher.

Unfortunately, the majority of the injured are children. Although fascinated by fireworks, children often underestimate their dangers. Even children aren't fast enough to move out of the way when a small bomb or a missile device doesn't go off as planned. Some states have legislation restricting fireworks that project or shoot. However, simple sparklers can cause injuries as well.

As midsummer approaches, suggest that the residents of your community avoid home fireworks and attend a public display instead.

Fall Safety

Changing Smoke Detector Batteries

About one-third of the smoke detectors in U.S. residences are inoperable, usually because of dead or missing batteries. This is a terrible statistic.

You cannot remind the citizenry often enough about the importance of checking and maintaining these proven lifesavers.

For years, the fire service has recommended that citizens note the autumnal changing of the clocks as a yearly reminder to change the batteries in their smoke detectors. The International Association of Fire Chiefs, in conjunction with a well-known battery maker, offers a campaign kit on this subject, available without charge to fire departments. The kit includes sample public service announcements that can be aired over your local broadcasting stations, also without charge. Additionally, the kit includes camera-ready artwork that the print media can use as written. Many larger stores in the area may capitalize on the event to hold special sales on such items as batteries, smoke detectors, fire extinguishers, and flashlights.

Fire Prevention Week

Many fire departments build their annual fire prevention activities around Fire Prevention Week in early October. This is their primary time to reach out to the public in a nonemergency situation. Whether your department regularly presents educational programs to the community or relies solely on Fire Prevention Week, the following suggestions will help you promote safety and add variety to your yearly activities.

First, however, a word about large-scale events. Such occasions require that many people put forth a colossal effort. You may think at first that your department can't provide the legwork and planning that are necessary, but consider some of the following options. The first, of course, is to combine your efforts with those of neighboring communities. Having the next town or two participate will not only make the event larger, it'll also increase the scale exponentially in comparison with the effort involved. Also, there'll be more people involved, thus spreading responsibility for the details. There are other personnel options to consider as well, such as using part-time or volunteer members, recruit classes, and youth groups. Our department has received excellent support from recruit classes. I contacted the coordinator of the local fire academy, and not only was I able to enlist adequate help, I was also able to expose the students to some of the precepts of fire safety education. Those students who volunteered received extra credit. Extend the invitation to the spouses and other relatives of the members, and you'll have quite a pool of workers from which to draw. Many family members of firefighters share the same commitment and enthusiasm and can be quite knowledgeable and helpful.

You can also visit Boy Scout and Girl Scout groups, high schools, and service clubs to enlist more volunteers. Senior citizens are often available to help with worthwhile projects, so head to some of their meeting places as well. If necessary, you can always advertise for helpers on the bulletin board of your local cable TV station.

Organize your volunteers early, and provide them with special helper badges or T-shirts as mementos. Try to make their experience a positive one, and rotate their assignments so there'll be some variety in the work they do for you. Thank your volunteers both publicly and privately. Recognize them and applaud them at the end of your event. Make up some extra souvenirs or certificates for them to take home. They'll have donated a piece of their time, so spend a little money and effort to let them know their work has been appreciated.

At the event, hand out literature that the visitors can take home for review. Handouts not only provide basic information, they also serve as reminders of what a person may have learned or experienced at the event itself. You can buy preprinted brochures or use a computer publishing program to make more personalized versions.

The following are some suggestions for attractions that you may wish to include in your Fire Prevention Week Open House.

Shooting Gallery: Have you ever tried to knock over a target at one of those shooting galleries at the county fair? Even if you didn't win a prize, you had fun trying. Set up a shooting gallery for the public, using a fire hose as a weapon. A $1\frac{1}{2}$-inch hose charged with hydrant pressure is usually sufficient to emphasize how heavy a hoseline can be, and the stream's reach is impressive to civilians. The cost of setting up this attraction is minimal. Use a beach ball set atop a traffic cone. Attach a three-foot string between the inflation valve of the ball and the cone. You'll have to cut the cone down halfway so the ball will sit atop it easily. This attraction requires two helpers—one to assist the public in handling the hose and the other to replace the ball each time it's knocked off. Usually you can assign a child to the latter task.

As an attraction, the flame house needn't be elaborate.

Flame House: This is an attraction similar to the shooting gallery. A flame house can be very simple, or you can make it quite elaborate. To construct one, you'll need three pieces of plywood, four feet square, hinged together to form the front and two sides, then painted to look like a house. Cut and paint a smaller, jagged piece of plywood to look like flame, then hinge this piece to the top or side so that it'll fall down backward when someone hits it with a stream of water. Attach a twenty-five-foot length of string or light line to the top of the flame. This attraction only requires a single operator, since he can both help the civilians handle the hoseline and reset the flame each time by pulling the string.

Pet Fire Dog: A firefighter in a dalmatian suit can provide numerous photo opportunities for your visitors. If budget isn't a problem and time is of the essence, the official NFPA mascot, Sparky, will show up for a cost of about $1,000. You can also find dalmatian Halloween costumes, but usually these are sized for children. Another option is to choose a different type of mascot. It should be an inherently fun figure so that it'll liven up the children. If possible, the mascot should relate to firefighting in some way.

In our community, we wanted to have a dalmatian for our mascot but didn't have a big budget. Our solution was to tap into the creativity of some of our own personnel. For less than a hundred dollars in materials, they constructed a dalmatian costume out of chicken wire, fiberglass, fur fabric, and a football helmet. Besides attending our Open Houses, this mascot has seen added service, delighting children at community events throughout the year.

To build such a costume, use the football helmet as the underlying structure of the mask. Frame out the head with chicken wire, then cover the form with several layers of fiberglass. Cut out holes for the eyes and nose, then stretch a layer of dark screening across them from the inside. You can obtain dark, see-through material from a fabric store. At the fabric store, you can also purchase spotted dalmation fur, enough to create gloves and ears, as well as to cover the head. For the rest of the costume, the firefighter may wear his turnout gear. Hope for a cool, windy day. Your safety officer will insist that your mascot be NFPA 1500 compliant in his garb.

Vehicle Extrication Demonstration: This is a definite crowd pleaser that involves very little cost. Contact your local towing company or junkyard to see whether it'll donate a vehicle with minor damage. Make sure the towing company will drop off the vehicle and pick it up again the day after your event. You may choose to place a mock victim in the car so that your responders can demonstrate their surveying, collaring, and backboarding techniques. With or without a victim, the crowd will enjoy watching your members cut the vehicle posts and flip the roof back on itself. This can also be a good training exercise for the firefighters involved, although advance practice is recommended. The team that conducts the extrication should be trained as a team before the event, and each member should be assigned a specific task for the length of the demonstration. Everything

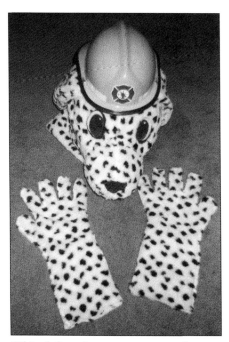

This dalmation costume cost less than a hundred dollars to make.

should be ready to go when you announce that the demonstration is about to begin.

Fire Extinguisher Demonstration: Attendees can participate in some hands-on learning when you teach adults about the proper use of fire extinguishers. You may want to check with your fire safety education division to see how this type of training is normally conducted.

Normally, you'll need a barrel or other receptacle in which to burn your demo fire. We use a fifty-five-gallon drum, cut about one to two feet up from the bottom. Paper works well as a fuel for starting the fire, but a small amount of liquid fuel may be needed for restarts after the flames have been put out a couple of times. After getting the audience in place, begin by mentioning that, in an emergency, they should always call 911 and ensure that the evacuation is nearly or fully complete before attempting to use a fire extinguisher. Briefly explain the classes of extinguishers, and recommend the ABC class for home use, since residences are subject to those three types of fires. Explain the PASS method for using an extinguisher: pull, aim, squeeze, and sweep.

Apparatus: Fire trucks, and lots of them, will make a breathtaking scene for your visitors as they approach the Open House. If possible, set up the ladders and towers to serve as landmarks for your guests. Fire truck manufacturers will often arrange to have a demo unit participate in your event. It's worth a call to see whether one is available.

Since many people are interested in seeing fire apparatus up close, you should have several members available who can answer questions and give impromptu tours of the equipment. Expect a myriad of questions. The children will want to know how tall the ladders are and where the hoses are stored. The adults will ask you about the engine, water-flow capability, and cost. For our Open Houses, we use an A-frame placard with this sort of information printed on it to help introduce the public to each vehicle. We also include basic stats, such as the year, make, model, and characteristics of any special equipment.

Additionally, check with your local contacts for the medevac helicopter that serves your area. Many of these companies will make special stops at fire department Open Houses, but you should book them early. First consider whether you have an adequate landing area for a helicopter and whether the visitors will have sufficient access to it.

Other agencies may wish to mount an exhibit. Local haz mat response agencies, high-angle rescue teams, urban search and rescue teams, and the like may all be available to you just for the asking. If your department doesn't provide transport services to the hospital, your audience may care to see an ambulance belonging to the provider that does. Call your local ambulance provider to arrange to have one of their vehicles present at your event.

Kitchen Fire Demonstration: It isn't necessary to build an entire kitchen to demonstrate how to extinguish a fire in one. Start with a pan and a lid. Explain that small pan fires are common, and describe how to keep the fire from spreading. The first and best action is to put a lid over the pan, making sure to protect your arm while doing so. Once the lid is covering the pan, turn off the burner. Explain how covering the pan removes oxygen from the system and turning off the flame removes heat. You may also wish to show how baking soda can kill the flames, or demonstrate a portable fire extinguisher rated for Class B fires.

Of course, the kitchen fire visuals would look more authentic if a stove were available. The stove doesn't have to work. A local waste management company may deliver one to your site, giving you a realistic prop at no cost.

If you can safely convert the stove to propane, you can use a

The essence of the kitchen fire demonstration is in using a lid to smother the flames, then stemming the source of heat.

portable propane tank to make the stove work. Our local gas company sent a representative to help us make such a conversion. Once we hooked up the tank, we were able to start real fires using cooking oil. Experiment a few times beforehand and see how long it takes to heat the oil to the point of ignition. Then, at the show, preheat the oil as appropriate before giving your demonstration.

If you want to go all out, build some kitchen scenery as a background to enclose the stove in a realistic setting. This will also create a barrier for keeping the audience at a safe distance while you conduct the demonstration. Make the walls three to four feet high, and keep them a few feet from the stove on all sides.

Firefighter Dress-Up: Many people, young and old, have dreamed of becoming a firefighter. Consider giving them the opportunity to suit up in turnout gear and have their picture taken. If you're using old clothing, have it cleaned before the event.

Everyone will want to wear a helmet. However, you should keep in mind two things. First, head lice are contagious, and second, at least one child suffered serious neck injury because the weight of the helmet was too much for his smaller muscles and bones. To avoid such problems, use plastic helmets as giveaways. You can find plastic firefighting helmets at party supply stores and elsewhere.

This is guaranteed to be a popular activity, so be ready with a few sets of gear and several workers. Park an apparatus nearby so all the photographers in attendance can use it as a backdrop. Don't be surprised if you're asked to get into the picture as well.

Stop, Drop, and Roll Practice: Borrow a gymnastic mat from one of your local schools or gymnastic centers so that children can practice this important skill. Emphasize the need to practice so that it'll be a reflex behavior in a time of crisis. Give positive comments to the youngsters as they successfully complete the sequence. Usually the younger children are willing to participate in this activity, but you'll have to use your powers of persuasion to get the older kids to try. Still, numerous lives have been saved by this technique.

Crawl Low Under Smoke Practice: Lay out a gymnastic mat on which children can practice this skill. Have two assistants hold an unfurled blanket at waist level to simulate smoke, then let the youngsters crawl under it and out to safety.

Another way to stage this attraction is to make a simple PVC pipe tunnel. It's easy and inexpensive to construct, and it eliminates the need for the two assistants. Use enough piping to create a tunnel at least five feet long. Attach the pieces together with tees, elbows, and an adhesive agent. Keep some of the connections unglued for breakdown and reassembly.

Normally, after the children successfully complete the crawl low under smoke and stop, drop, and roll exercises, we give them a prize, such as trading cards or a small toy of some kind.

Coloring Contest: Use a coloringbook-style outline as the basis of a contest. Establish clear-cut entry rules, and stick to them. The contest literature should include the age groups allowed to participate, the prizes to be awarded, the judging criteria, the entry deadline, and the contest date. One easy way to handle this is to draw the winners randomly, regardless of the quality of the picture. Break the entrants into appropriate age groups, and select a winner from each. Using this approach helps you avoid any hint of favoritism.

The winner can be awarded a variety of prizes, including free meals at local restaurants, coupons for local stores, or some other gift. In my community, the grand prize winner of each age category becomes Firefighter for a Day. The winners are picked up at school in a fire truck, driven to a local restaurant for lunch, then returned to school afterward. This proves to be a memorable event for the winners, as well as an honor for the school. The winner is given an escort back to class by firefighters, who then discuss the contest, the Open House, and topics of fire safety. This event has received much attention in our community. It has proved to be a positive, upbeat story welcomed by the print and broadcast media alike, so don't forget to invite the local reporters.

Refreshments: Local bakeries may donate treats for your audience, or perhaps a local service group or scouting den may volunteer to come in to sell food and drinks. You'll be too busy to spend a lot of time dealing with edible concessions, so have some outside organization handle it for you.

Feedback: Hand out evaluation sheets to the workers immediately after the Open House so that you can elicit their comments while the event is still fresh in their minds. Find out their thoughts as to which activities went well, which need work, and how to improve the Open House for next year.

Halloween Safety

Children await Halloween with great anticipation. Many aren't aware of the inherent dangers of walking or running in costumes at night, often in the presence of traffic.

You can place the following Halloween safety tips in local newspapers, newsletters, and advertisements, and even broadcast them by means of radio, TV, and cable.

- Look for costumes, masks, beards, and wigs that are labeled as being flame resistant. Although this doesn't mean that the items won't catch fire, it does mean that they will resist burning and should extinguish quickly once you remove them from the ignition source. Avoid using flimsy materials and outfits with big, baggy sleeves or billowy skirts, since these can be hazardous near candles and other ignition sources.
- Costumes should be bright enough to be clearly visible to motorists at night. For greater visibility in twilight or darkness, trim the costumes with reflective tape so that they'll glow in a headlight beam. Bags and sacks should also be light in color and decorated with reflective tape. You can buy reflective tape in hardware, bicycle, and sporting goods stores. Children should carry flashlights so they can see and be seen more easily.
- Costumes should be short enough to prevent children from tripping on them. Children should wear sturdy, well-fitting shoes. Mom's high heels aren't a great idea for trooping around in the dark.
- Tie hats and scarves securely to prevent them from slipping over the children's faces and ears.
- Since loose-fitting masks can restrict breathing and obscure vision, consider creating a mask with cosmetics and face paint instead. If the child does wear a mask, make sure that it fits securely and has eyeholes large enough to allow full vision.
- All swords, knives, and other such accessories should be of a soft, pliant material.
- An older, responsible child or adult should always accompany smaller children. Children should use the sidewalk instead of the street, and they should walk, not run, from house to house. Caution them against running out from between parked cars or across other people's yards, where ornaments, furniture, and clotheslines could present hazards.
- Children should only go to homes where the residents have an outside light on as a sign of welcome. They should never enter a home or an apartment unless accompanied by an adult whom they know.
- Those who receive trick or treaters should remove anything that could be an obstacle or other hazard. Keep candlelit jack-o'-lanterns away from landings and doorsteps, where costumes could brush against the flame. Better still, illuminate the jack-o'-lantern with a flashlight. Keep

indoor jack-o'-lanterns away from curtains, decorations, and other furnishings that could ignite.
- Children should not eat any treats before they get home.
- Parents and guardians should examine all treats carefully in bright lighting for evidence of tampering before allowing children to eat them.

Winter Safety

The risk of fire normally increases during the year-end holiday season. Whether from the increased use of decorations, the abundance of parties, the extra burden on home heating equipment, or some other factors, the holiday season spells disaster to many families each year. Encourage the citizens of your community to give some thought to fire safety and fire prevention, for this can go a long way toward ensuring a safe and happy holiday.

Christmas Trees

- Check the Christmas tree for freshness before purchasing it. One way to do this is to rub it gently. If the needles fall off, the tree is too dry. Fresh needles will bend between your fingers without breaking. Don't rely on color to tell. The tree may be sprayed green to fool you. If you don't cut the tree yourself, make another angled cut at its base when you get it home to allow for better water absorption.
- In placing the tree, choose an area that isn't blocking any exits and that is away from heat sources.
- Carefully place the tree in a stand with water. Check the water level daily. Replace the water as necessary so the tree won't dry out prematurely.
- Take the tree down shortly after Christmas.
- The Environmental Protection Agency (EPA) has designed a recipe to help increase a tree's uptake of fluid. Combine the following ingredients in a two-gallon bucket:

  - 1 gallon hot water.
  - 1 pint clear Karo syrup.

- 4 ounces liquid chlorine bleach.
- 1/2 teaspoon Borax.
- 2 ounces cider vinegar.
- 2 ounces of wetting agent or liquid Woolite®.

Fill the remainder of the bucket with hot water, and place the trunk of the tree in the bucket. For best results, leave the tree in the bucket for five days before bringing it inside. Use the remaining liquid mixture to fill the reservoir of the tree stand.

- Metal trees can become serious electrical shock hazards if they're frayed or if they're decorated with damaged lights. Any bare wire that touches the metal will energize the whole tree. Carefully inspect the lights before using them.
- Plastic trees should be made of fire-resistant material and labeled as such. This only means that they won't catch fire easily, but they will burn. Therefore, take all the precautions with this type of tree as you would with a real tree.

Holiday Lighting

- Be sure that all indoor and outdoor decorative lighting has been tested and labeled by an independent testing laboratory, such as Underwriters Laboratories.
- Before each season, inspect the cords for cracks or frayed areas. Other signs that the lights need to be replaced are loose connections, broken sockets, and bare wires. Never attempt to repair such faults with electrical tape. It is always safest to replace lights as necessary. Test them for a while under constant supervision before putting them up as decorations.
- Keep outdoor lighting and electrical connections off the ground, and have sockets hang down so that water won't seep into them. Cover any connections with plastic wrap, and seal the ends with electrical tape.
- Don't run cords under rugs, where they can be hidden from inspection. Walking on concealed wiring causes extra wear, which can lead to a fire.
- Unplug all holiday lights before leaving or going to bed.

Candles

• Allow sufficient room between candles and combustible items. Pay particular attention to areas where you don't normally put candles, such as windowsills, shelves, desks, and the like.

• Don't leave candles unattended.

• Always blow out any burning candles before going to sleep.

• Keep children at a distance, and especially don't allow them to play with the molten wax or flame.

Fireplaces

• Children and fires don't mix. Keep children well away from open flames, and don't leave the fireplace unattended. If a youngster is curious about fire, give him a helper job, such as gathering kindling or bringing in wood for the fireplace.

• Inspect chimneys annually, and have them cleaned as necessary.

• Avoid burning trash or wrapping paper, since this will lead to a premature buildup of creosote, which will block chimney ventilation and pose a fire hazard.

• Keep fireplace screens in place during use, and keep any combustibles at least three feet away.

• Do not leave the house while a fire is burning in the fireplace, and ensure that the fire is completely out before going to bed.

• If a fire breaks out in the chimney, always call the fire department before attempting to put it out yourself with a fire extinguisher.

Portable Space Heaters

• Portable space heaters need space around them. Leave an area at least three feet clear in all directions. These heaters should only be used for their intended purpose.

• Many fires are started by blankets or clothes draped on top of a space heater.

• If the unit runs on fuel, only use the proper fuel. Refill the heater only in ventilated areas and when the unit is cool.

• Vent the heater as outlined in the manufacturer's guidelines.

- When leaving the room or going to bed at night, shut off portable space heaters.

Smoke Detectors

- What more need be said? The increased hazards associated with the holidays point out all the more the importance of having operational smoke detectors in your home. A working smoke detector in your home lowers by half your chances of dying in a fire. Test the unit monthly, replace the battery annually, and dust the inside when you change the battery. Stuff one into a stocking and give it to a loved one.

Houseguests

- If out-of-town guests are staying in your home, include them in the family escape plan, and practice it with them for their benefit as well as your own. Let them know where your smoke detectors are located and what they sound like. Also, take the time to show them how to open the windows in the house. Show them the best means of escape and where to meet outside in the event of a fire.

Chapter 11

Preventing a Poisoning

Targeting Children

Poisonings are a major concern in the United States. Children under the age of five are most often its victims. Target your poisoning safety information to those groups most likely to be responsible for children, such as parents, baby-sitters, and day care centers. This topic should be part of your other safety programs as well.

In addition, you may want to organize a special campaign during National Poison Prevention Week, the third week in March. You may promote your program by all the usual means of broadcast and print media.

Our department has constructed a banner-type display for this event. The top bears the legend "Children Act Fast ... So Do Poisons," and the bottom says, "Beware of Poison Look-Alikes." Framed between these two slogans, glued to narrow shelves and shielded behind a sheet of acrylic, is an assortment of candy with their look-alike medicines beside them. This sign serves as a reminder to adults that they should keep even nonprescription medicines away from children, who do not know that they can be harmful.

The following are some of the key points to make during any presentation about poisoning:

- Keep children under constant supervision.
- Store prescription medications in pill-minders and out of the reach of children.
- Keep over-the-counter preparations, beauty aids, and housecleaning products in a safe place, out of the reach of children. Even toothpastes and mouthwashes, especially if they contain alcohol, can harm children.
- Store all medicines separately from household products, and store all household chemicals away from food.
- Keep items in their original containers. Do not remove the labels from the container.
- Use products with child-resistant packaging, and secure the container after use.
- Don't take medicine in front of children, because kids characteristically imitate grown-ups.
- Never call medicine "candy."
- To avoid mix-ups, never dispense medicine in the dark. Always turn on a light when taking medicine yourself or giving it to others.
- Periodically clean out the medicine cabinets.
- Keep the number of the nearest Poison Control Center on or near your telephone.
- Keep a bottle of syrup of ipecac available. Be prepared to use it on the advice of a physician.
- If you believe a child has ingested a poisonous substance, remain calm and call 911 first, then the Poison Control Center or a physician. If you feel that the poisoning isn't life-threatening, call the local Poison Control Center first. It is open twenty-four hours a day. The Poison Control Center estimates that about 85 percent of the phone calls it receives for the ingestion of a poisonous substance do not require any other intervention.
- When calling the Poison Control Center, be prepared to provide the following information: (1) the victim's age, (2) the victim's weight, (3) any pre-existing health conditions or problems, (4) the type of poison involved and the manner by which it entered the body, whether through ingestion, inhalation, absorption, splashing, or injection, (5) the first aid measures administered, (6) whether the person has vomited, (7) your location and how long it might take to get to the hospital, and (8) the mental status of the victim.

Chapter 12

Public Service Announcements

The Succinct Memory Aid

A public service announcement, or PSA, is designed to act as a quick reminder to an audience, given by way of a specific medium. By law, radio stations are required to make a certain amount of time available for PSAs and other informational purposes. Television and cable stations must also make time available. Sometimes a given station will offer airtime at a reduced cost, but time slots are also available without charge.

Since the intention of a PSA is to jog a memory from time to time, each such message usually covers only one or two basic points. Television and cable stations might add fire footage and use a voiceover format to advance the message, or they may simply print the announcement on the screen to be read by viewers. Newspapers can publish a sidebar featuring safety tips to supplement an article related to a fire or other emergency.

The following PSAs can be used alone, or they may be combined. You can also create your own. Getting a PSA out onto the airwaves usually takes just a telephone call or two, and you can target the message as appropriate, given the day or season of the year.

Sample Ten-Second PSAs

• When you set your clock back this fall, change the batteries in your smoke detectors, too. This simple act can help double your chances of surviving a fire in your home. A reminder to change your clock and your battery from the Hometown Fire Department and WXYZ radio.

• When a child strikes a match, fire strikes back. Don't let it strike your family. The Hometown Fire Department and radio station WXYZ remind you to keep matches and lighters in a safe place.

• Another fire prevention tip from the Hometown Fire Department and radio station WXYZ. Install smoke detectors inside and outside of bedrooms to give all members of your family plenty of time to get out alive.

• Plan your home escape route in case of fire. Make sure everyone in your family knows two ways out of every room and where the family meeting place is. Then, practice the plan. This has been a reminder from the Hometown Fire Department and WXYZ radio.

• Can you find the fire exits in your home in the dark? Do you and all in your house know where the nearest outside phone is? Make sure you know. Learn not to burn. This reminder is brought to you by the Hometown Fire Department and WXYZ radio.

• Keep your children fire safe. Don't let them burn. Practice the stop, drop, and roll technique with them, and make sure they know when to use it. This message provided by the Hometown Fire Department and WXYZ radio.

Sample Fifteen-Second PSAs

• Your house is on fire. You know you have to get out fast. But there's one thing you want to save—an old photo, legal documents, or a family heirloom. Should you chance it? No. Remember, if you're in a fire, there's no time. If you try to save one thing, you could lose one thing—your life. This is fire! This has been a public service announcement from the Hometown Fire Department and WXYZ radio.

• Stop what you're doing. Close your eyes. Spin around. Sound like a child's game? It's not. It's what you'll feel like in a fire. Blind. Disoriented. Confused. Remember: If a smoke detector goes off tonight, you'll wake up to total darkness. Fire is dark, not bright. This is fire. This has been a public service announcement from the Hometown Fire Department and WXYZ radio.

• If there's a fire in your house tonight, would you wake up? Chances are, you wouldn't. Gases and smoke lull you into deeper sleep. You pass out before the fire reaches your door. Remember: You won't wake up to the noise or burning smell of fire because deadly smoke and gases can get you first. This is fire! This has been a public service announcement from the Hometown Fire Department and WXYZ radio.

• How hot can a fire get? So hot that your pajamas melt into your skin. So hot that the air scorches your lungs. Fire can get so hot that everything explodes—from your TV to your armchair. Remember: In a fire, crawl on the floor, where the air isn't as hot. Heat from a fire can kill. This is fire. This has been a public service announcement from the Hometown Fire Department and WXYZ radio.

• Smoke detectors. You know, they don't cost much—about ten dollars. They're easy to put up. The fire department may even install them for you. And smoke detectors can save your life. This has been a public service announcement from the Hometown Fire Department and WXYZ radio.

• Smoking and gasoline never mix. So think about it the next time you're filling your lawn mower, gas can, or automobile. Don't smoke. Just the fumes from the gasoline can ignite and cause a fire. And be sure to fill your gasoline-powered engines outside, just in case a fire does start. This has been a public service announcement from the Hometown Fire Department and WXYZ radio.

- The fumes from a can of gasoline can explode under the right conditions. You should always store gasoline in a proper container—never in glass. And store it away from areas such as a basement or furnace room. Whenever you use gasoline, be sure to do so outside, where the fumes can disperse into the air. And remember—never smoke around gasoline. This has been a public service announcement of the Hometown Fire Department and WXYZ radio.

- Did you know that one-third of the people who have smoke detectors don't have working batteries in them? Help prevent a needless tragedy. This fall, when you set your clock back, change the batteries in your smoke detectors, too. Make this change a meaningful one. A reminder from the Hometown Fire Department and WXYZ radio.

Sample Thirty-Second PSAs

- A fire starts in your house late at night. In thirty seconds, a dropped match can become a fire burning out of control. The heat is intense. The air scorches your lungs. The fire's black smoke gushes through your house. It's so dark, you can't find your way through your own room. The toxic gases in the smoke put your family into a deeper sleep. None of you wake up to hear the fire raging outside your doors. Only minutes have gone by since the fire started, and it's already too late to escape. Fire—fast, hot, dark, and deadly! Understand it. Prepare for it. This is fire! This has been a public service announcement from the Hometown Fire Department and WXYZ radio.

- For older Americans, fire in the home often means injury or even death. Most of these fires don't have to happen. Unfortunately, some people just aren't aware of the dangers. The fire hazards that threaten senior citizens the most—kitchen fires and heater fires—can be prevented. A little more caution around the stove, keeping curtains and clothing away from heaters—simple things like these can help prevent a fire. So if you're over sixty-five, or even if you're not, make fire safety a priority. This has been a public service announcement from the Hometown Fire Department and WXYZ radio.

- The countertop area of a kitchen is an area where many accidents happen. You can help prevent these accidents by observing a few rules. First, never let appliance cords hang over the counter edge, where little children can grab them and pull the appliance down on top of themselves. Second, when cooking, don't wear loose-fitting clothes that could catch fire by coming in contact with burners. Third, keep hot liquids and knives back away from the counter edge, where small hands can't reach them. This has been a public service announcement from the Hometown Fire Department and WXYZ radio.

- According to the U.S. Fire Administration, smoke detectors decrease your risk of dying in a home fire by as much as fifty percent. The earlier the fire is discovered, the better your chances of survival. Smoke detectors should be on every level of your home. Test your detector at least once a month, and change the batteries once a year. A chirping noise means the detector needs a new battery. It's a real protector, the smoke detector. This has been a public service announcement from the Hometown Fire Department and WXYZ radio.

- Most fatal fires occur at night, when you and your family are asleep. When seconds can make the difference between life and death. Are you protected? If you have one or more working smoke detectors in your home, you double your chances of surviving a fire. That's protection! And smoke detectors are inexpensive and easy to install. Smoke detectors—they're real protectors. This has been a public service announcement from the Hometown Fire Department and WXYZ radio.

- Here's a reminder from the U.S. Fire Administration. Having a working smoke detector on every level in your home doubles your chances of surviving a fire. It is estimated that out of all the smoke detectors in the U.S. today, one-third to one-half aren't maintained or have been disabled. Clean and test each smoke detector at least once a month, and change the battery yearly. It's a real protector, the smoke detector. This has been a public service announcement from the Hometown Fire Department and WXYZ radio.

- Most fatal fires occur at night, when you and your family are asleep. Smoke detectors double your chances of surviving a fire, but they must be properly maintained. Smoke detectors should be tested monthly and the battery changed at least once a year—on your birthday, for example. And remember, never disconnect the battery. Smoke detectors—they're real protectors. This has been a public service announcement from the Hometown Fire Department and WXYZ radio.

- This is a test. It's multiple choice. Ready? In a medical emergency, what should you do and say when you call for help? Should you (A) dial the number and scream hysterically, (B) ask if you can be on TV, or (C) stay calm and tell the emergency medical services operator who is hurt or sick, what is wrong, and where to find the victim? If you said (C), tell the EMS dispatcher who, what, and where, you made the right choice. So remember, who, what, and where when you call EMS for emergency medical care. This has been a public service announcement from the Hometown Fire Department and WXYZ radio.

- This is a test. It's multiple choice. Ready? If something goes wrong, when should you call emergency medical services for help? (A) When your sink is clogged, (B) when the lights go out, or (C) when someone is seriously hurt or sick? If you said (C), call EMS when someone is hurt or sick, you made the right call. Call emergency medical services in the event of an actual emergency. Remember, if it's medical, call EMS. If it's a clogged sink, call the plumber. This has been a public service announcement from the Hometown Fire Department and WXYZ radio.

- You probably have at least one emergency lifesaving device in your home already. You can learn how to use it, right now, in two easy steps. In an emergency, simply pick up your phone and dial 911. That's it, 911. Call 911 for help in all kinds of medical, police, and fire emergencies. When you need help fast, your phone can be a real lifesaver. It's as easy as 911. Dial 911 in emergencies, and make the right call. This has been a public service announcement from the Hometown Fire Department and WXYZ radio.

Public Service Announcements

• The Hometown Fire Department issues an urgent call to all residents this weekend. Take a moment to change the batteries in your smoke detector. It's one of the simplest, easiest things you can do—and it could save your life. Nationwide, one-third of all homes with smoke detectors don't have working batteries. So please, when you change your clock, change the batteries in your smoke detectors. A reminder from the Hometown Fire Department and WXYZ radio.

• When a child strikes a match, fire strikes back. Don't let it strike your family. Teach your child that fire is no toy. It takes only about two minutes for the flame from a single match to set an entire room on fire and less than five minutes for that fire to overtake an entire house. Find a safe place for matches and lighters—far away from curious fingers. This station and the Hometown Fire Department want to remind you and your children that fire strikes back!

• When a child strikes a match, fire strikes back. Don't let it strike your family. Nearly twenty-five percent of the fires that kill young children are started by the children themselves. At home, curious kids usually play with fire in the bedroom, where there are lots of things that catch fire easily. Teach your children that fire is no toy. Keep matches and lighters in a safe place. The Hometown Fire Department and radio station WXYZ remind you and your children that fire strikes back!

• Clean it up before it burns up! The Hometown Fire Department and WXYZ radio want you to know that trash causes more than one hundred home fires each day. Help protect your home. Clean the basements, garages, and sheds. Get rid of old magazines, newspapers, clothing, and other unused items. Store gasoline and other flammable liquids outside the home in approved containers. Remember, trash and clutter give fire a place to start. Clean up today so you'll be safe tomorrow.

• Many children care for themselves after school while their parents are away. All children should learn fire safety precautions, but these latchkey children especially need to know how to protect themselves from fire early on. The Hometown Fire Department and WXYZ radio want to remind parents and children to be extremely careful when cooking. If clothing catches fire, immediately perform the stop, drop, and roll technique, and if you get burned, cool the burn with water. Remember, when a fire starts, it's too late to practice.

• Good grooming is important, but not at the expense of safety. The Hometown Fire Department and WXYZ radio remind you to use all beauty and health care appliances safely. Keep anything electrical away from water. Never dip electric styling equipment in water to dampen hair. Keep air vents on hair dryers open to prevent overheating, and never leave any heat-producing appliance unattended. If you don't have ground fault interrupters, consider installing them. Be beautiful, but be safe as well.

• Most fatal fires occur at night, so a smoke detector has to smell the fire while you sleep and warn you so you can escape safely. The Hometown Fire Department and WXYZ radio want to remind you to test your smoke detectors every month. If your detectors are wired in, have battery backups in case the electricity goes out. Change the batteries at least once a year. Every household should have and practice a home fire escape plan that includes at least two ways out of each room. Don't leave your life to chance.

• The Hometown Fire Department reminds motorists that the law requires you to pull to the right-hand curb and allow emergency vehicles to pass. Failure to do so may cause serious accidents or delays in an emergency. Move over to the right to allow the vehicle through. If you find yourself behind an emergency vehicle, keep well behind it in case of sudden stops. Remember, pull to the right for sirens and lights.

• Today would be a good day to test your smoke detectors to be sure they're working. If they aren't, try replacing the batteries with new ones to be certain your family is protected. And practice what you would do if there were a fire in your home. Get low, get out, and stay out. Meet your family at a predetermined meeting place outside, and have a family member call 911 from a neighbor's phone. This message is brought to you by the Hometown Fire Department and WXYZ radio.

• If a fire were to break out in your home tonight while you were sleeping, would you and your family get out alive? Develop a home fire escape plan, and practice it on a regular basis. Working smoke detectors can reduce by half your risk of dying in a fire. And remember, if there is a fire, get low, get out, and stay out. Go straight to the family meeting place. This message is brought to you by the Hometown Fire Department and WXYZ radio.

Sample Forty-Five-Second PSAs

• The Hometown Fire Department has a special message for parents. Most fatal fires occur at night, when you and your family are sleeping, and when seconds can mean the difference between life and death. Smoke detectors, when properly installed and maintained, double your chances of surviving a fire. Smoke detectors are inexpensive. Put at least one on every level of your home. Remember to test your smoke detectors monthly and to change the battery at least once a year—on your child's birthday, Christmas, or some other holiday—because you shouldn't leave your family's safety to chance. Smoke detectors—they're real protectors.

• Smoke detectors. Twenty-five years ago, most people had never heard of them. Today, nearly two-thirds of American homes are protected by them. If your home doesn't have at least one smoke detector, it should. The Hometown Fire Department and WXYZ radio want to remind you that most fatal fires happen late at night, when most people are sleeping. Poisonous smoke creeps through the house and kills before the victims have a chance to wake up. Don't let it happen to you. Make sure your home has smoke detectors. Clean and test them every month, and change the batteries every year. And make sure you know what to do if there is a fire. Call your local fire department for advice. We want you to sleep safely every night.

• It's the time of year when cool outside temperatures cause us to turn on our heating systems. They make our homes warm and cozy, but they're also the leading causes of fires. Have all heating equipment, including chimneys, serviced every year by reputable professionals. Keep anything that can burn at least three feet away from any fuels. Use portable heaters only if they carry the UL label and only if they have sturdy cords. Never use them with an extension cord. Fill kerosene heaters outside, and use them only in well-ventilated rooms. Never leave portable heaters unattended. Be careful while keeping your family warm and safe. This is a reminder from the Hometown Fire Department and WXYZ radio.

• We know you always try to be careful. But just suppose a fire breaks out in your home late at night while everyone is asleep. Does everyone in the home know what to do? Would they roll out of bed, crawling low to keep below the poisonous gases that are swirling near the ceiling? Do they know that a hot door means the fire is right beyond it, so they should choose another way out of the room? Have you designated a family meeting place so you'll know that everyone is safe? If your answer to any of these questions wasn't a definite yes, don't take any more chances. Sit down together tonight, make your exit plan, and practice it. This is a reminder from the Hometown Fire Department and WXYZ radio.

• Often we don't even notice them, but exit signs can help prevent tragedy in an emergency. You probably know more than one way to escape from your home in case of fire, but what about your favorite restaurant, grocery store, movie theater, sports arena, or public office building? When you're in any public building, note where the exits are. In an emergency, if everyone tries to get out the same door, crowding, pushing, and panic can result. People can be hurt and even killed. Make it a habit to check for exits, and teach your children to do the same. The Hometown Fire Department and WXYZ radio want you to be fire safe!

• Tired of giving the same old holiday gifts year after year? Can't face the thought of yet another pair of slippers or a candy dish? How about gifts that show you really care? The Hometown Fire Department and WXYZ radio suggest you give smoke detectors, portable fire extinguishers, or home escape ladders to your loved ones. Thousands of Americans die each year in home fires, many during the holiday season. Do something about it this year. For tips on what to look for in buying any of these items, call your local fire department. This holiday season, give a gift that can save lives, and have a safe and happy holiday.

- It's hard to beat the flavor of burgers, steaks, and corn grilled on the barbecue. However, the Hometown Fire Department and WXYZ radio want you to use some common sense and cook safely. If you use lighter fluid, use it sparingly, and never on a fire that's already started to burn. Keep away from overhanging trees, awnings, and roofs. Remember to keep a bucket of water or a garden hose nearby in case of a sudden flare-up. And use long-handled outdoor cooking tools to keep your hands safely away from the grill and coals. Keep children away from the grill, and never leave it burning unattended. Enjoy the nice weather barbecues, but do so safely.

- Thousands of people are seriously injured every year by cooking fires. The Hometown Fire Department and WXYZ radio suggest you keep your kitchen fire safe and know what to do if a fire does start. Keep cooking surfaces clean so grease can't build up and start a fire. Never leaving cooking food unattended. If a fire starts in a pan on the stove, slide the lid on top of the pan to smother the fire. If the pot is in the oven, turn off the heat and keep the oven door closed. Or, you can use a fire extinguisher if you have one and know how to use it. If the fire's too big, don't try to fight it. Leave the house right away, and call the fire department from a neighbor's phone. No property is worth a risk to your life.

- In an emergency, minutes can seem like hours. If you call 911, we will be able to come to your aid more quickly if you help show us the way. Please give the dispatcher as much information as possible so we can locate you and determine the type of assistance you need. Once the dispatcher has the information and instructs you to hang up, go or have someone else go outside and wait for the responding emergency vehicles and assist them in locating the problem. Another way you can help is by having your street numbers posted so that they're highly visible. They should be at least three inches high and either reflective or in a contrasting color. Remember, if you need assistance, help show us the way. This has been a reminder from the Hometown Fire Department and WXYZ radio.

• Fire is hot, deadly, and fast. It only takes a minute for a fire burning in a house to grow to about twice its original size. If you're the one to discover a fire, the Hometown Fire Department and WXYZ radio remind you to do the following. Sound the alarm to let others in the building know they need to evacuate. As you leave, close as many doors as possible behind you to help confine the fire. Call the fire department from a remote phone in a safe area. Don't take a chance on getting trapped. Make sure no one goes back into the building. Take a count of all people who are normally in the building to see whether everyone is out safely. Report the findings to the first-arriving fire department crew. Remember, fire is hot, deadly, and fast.

Sample Sixty-Second PSAs

• Cold weather is coming soon, and a fire in the fireplace is going to feel good. But it wouldn't feel too good if that fire were burning down your house and furnishings. And if you fail to take the proper precautions, that could happen! So have your furnace, chimneys, and flues inspected and cleaned, if necessary, before the time comes to use them. See that any portable space heaters you own are clean and in good repair before you turn them on. Go over the rules for space heaters with the family. Keep combustibles three feet away, and never leave space heaters on when the room is unoccupied. Provide a screen for your fireplace, and see that it is always in place whenever you light an open fire. Keep materials away from the fire, and make supervision by an adult a priority. Make sure that all the fires that burn in your home are friendly fires. This has been a reminder from the Hometown Fire Department and WXYZ radio.

• Are you planning a holiday gathering in your home, office, house of worship, or school? Holiday parties often center around a Christmas tree, and the Hometown Fire Department and WXYZ radio remind you that evergreen trees can catch fire easily and burn rapidly. So make safety as much a part of your party plans as the decorations and refreshments. Take the following precautions for a fire-safe gathering. First, don't set the tree near a stairway or elevator shaft, which would produce a draft. Don't let it block a door or any exit. Provide plenty of ashtrays for smokers, and don't allow smoking near the tree. See that all decorations in the room have been approved as flameproof. Keep the tree well watered, and take it down as soon as it shows signs of drying out. Don't let fire come to your party!

• Christmas is coming, bringing with it a serious home fire hazard—the Christmas tree. Evergreen trees ignite easily and burn rapidly, so for safety's sake, place your tree in a bucket of water before bringing it inside. When you bring it in, make a fresh cut just above the existing cut to help water absorption, and keep your tree away from possible sources of heat or sparks, such as heaters, fireplaces, electricity, and open lights. Use only flame-resistant decorations, and be sure that all lighting sets and electrical cords are in good condition and have the UL label. While your tree is up, keep it well watered, and periodically check to see whether its needles are drying out and falling down. When they do, it's time to take the tree down and discard it—outside. The Hometown Fire Department and WXYZ radio want to keep your holiday season a fire-safe and happy one!

Sample Short, Printed PSAs

In many cases, your local cable company may have a screen with numerous printed messages that run continuously throughout the day and late at night. Usually the station will be more than happy to run fire safety PSAs along with other messages. The screens are small, and having a message go beyond one screen becomes confusing for the viewer. Try to keep the message short and to the point. Consider the following examples as starters:

• Develop and practice your home fire escape plan. This includes having two ways out of every room in the house, establishing an outside meeting place, and checking smoke detectors regularly.

• Your chances of dying in a fire are cut in half by having operational smoke detectors in your home. Test your detectors once a month, and replace the batteries once a year.

• Practice makes perfect. By practicing the stop, drop, and roll procedure to snuff fire out of clothing, you'll increase your chances of remembering what to do in an emergency.

• Remember that smoke and heat rise in a burning building. If you are in a smoky building, remember to crawl low under smoke.

• If you aren't sure of how to develop a home fire escape plan, stop by your local fire station, or check out our video at fire headquarters. Viewing the tape is free.

• When cooking on the stove, always keep the lid handy. If a fire breaks out in the pan, slide the lid over the top of the pan, and turn off the heat to the unit.

• It's a real protector, the smoke detector. Don't stay home without one, and test yours every month.

• If your smoke detectors sound in the middle of the night, don't hide. Go outside. Crawl out of the home, and meet at your family meeting place.

• Fire causes an injury every eighteen minutes. Practice fire safety every day. If you do get burned, flush the area with cool water.

• Fire kills a U.S. citizen every one hundred minutes. Develop a home fire escape plan, and practice it.

• A fire is reported to a U.S. fire department every sixteen seconds. Heating equipment is a major cause. Watch what you heat!

• Most home fires begin in the kitchen. Put a lid on grease fires, turn off the heat to the stove, and call the fire department.

• Don't wait until fire strikes to plan your escape. Do it now. Know two ways out of every room in the home, and designate an outside meeting place for your family.

- Children under the age of five have a fire death rate of more than twice the national average. Keep matches and lighters in the right hands.

- Thousands of Americans die every year in fires, most of them in their own homes. Plan your home fire escape plan, and practice it before fire strikes.

Computer-Animated PSAs

No doubt you've seen the marvels of computer animation, but you may not yet know how to create one for your own purposes. Check your local yellow pages under "animation," or consult a local producer (possibly through your cable company) to determine the feasibility of putting an animated public service announcement on the air.

We asked a local computer animation company to help us come up with new, high-tech ways to get the message of fire safety out to the public. We were lucky. We got much more than we expected. The producer was kind enough to design, produce, and edit a forty-five second PSA about the benefits of stop, drop, and roll when clothing catches fire.

In our video, a man is barbecuing outside when his clothing catches fire. With the help of an announcer and his loyal dog, the man is able to snuff out the fire. This PSA is used as a stand-alone message on our local cable channel. Its approach emphasizes learning without highlighting negative consequences.

Try animation. The visual interest of the imagery will help capture the attention of your audience.

Other Programming

Another option is to use expanded programming to get your messages across. The opportunities afforded by the modern media, particularly cable, are so vast that you shouldn't necessarily limit your department's voice to PSAs alone.

You should consult the cable company representative to see whether fire department activities might appear on nonfire programming. For example, many communities offer biweekly programs of events in the area. In convincing the cable company to include your department's event in such a show, describe the theme and occasion of the event; when and where it'll be; and how long it'll last. Whet their appetites by including a short demonstration of a topic that you intend to cover.

A bigger step to take is to consider airing a fire department program. Such a show can be produced solely by your department or in conjunction with the cable company. Programs that cover different safety issues, training, new apparatus, EMS activities, and interviews with department members have been successfully produced in many communities.

Having been involved in a number of these shows, I offer the following recommendations. First, involve as many members as possible to maintain interest within the department. You should brainstorm topics to be covered with the cable representative before beginning the program outline. After settling on the subject matter, confine the length of the program to fifteen minutes, at least initially. Also, keep each show related to one overall theme. A program on EMS, for example, might include a video tour of a vehicle; an explanation of how to notify EMS; how dispatchers handle a typical call; how the fire station receives that call; how drivers are supposed to react to lights and sirens; and possibly a mock run of the ambulance.

As additional material, you can review upcoming training sessions, keeping newsworthy items in mind. Give as much information as possible about upcoming events, but include pieces about personal achievements as well, not just departmental matters. The cable company may wish to show the more human side of the fire department from time to time.

Chapter 13

Print Messages

Grocery Bags

People need reminders. You may have a number of effective programs going on, but people need to be reminded to practice what they've learned. Fire departments do this through radio, television, and cable PSAs, as well as brochures, stickers, and a variety of other innovative ways. How about adding grocery bags to the list?

Approach the store owner or manager with the idea. All our department had to do was supply the owners with the artwork and a short message, and they did the rest.

This project can be enhanced by adding high school or college art students to your team. Give the teacher an idea of what you want, and the class will work on the details.

The more ways that you can provide fire and life safety messages to your community, the greater the chance that the community will respond to those messages.

Newspaper Columns

There are newspaper editors who'll listen to what the fire department has to say, then write a column that captures the essence of the interview.

Grocery bags provide a viable way of getting out the fire safety message at the community level.

Departments in such a locality are lucky. Other departments try to convince their local journalists to cover their stories but never receive the cooperation they seek. If such is the case in your area, try to come up with ways to get coverage of the fire and life safety information. This may mean offering to write the columns yourself. You may or may not be given a byline.

Newsletters

Most schools publish newsletters. Meet with the principal initially to get the go-ahead, then periodically drop off items such as clip art, PSAs, or promotional material. Keep the items short and to the point, since space is usually limited.

Some PTAs have separate newspapers that go out to parents. They usually have a mailbox at most school offices, so drop off copies of your material to them as well. Preschools also have parent newsletters. Stop by and ask for help in getting out your message. In the past, our department has attached its Open House flyers to these newsletters. In some cases, we shortened the information enough to insert it in the newsletter itself. Houses of worship are another newsletter outlet. Usually they have room for brief safety items. Public service announcements and short pieces are usually best.

In general, keep an eye out for others who publish newsletters in your community. Banks, credit unions, real estate offices, the local chamber of commerce, service clubs, veterans associations, larger businesses, hospitals, and nursing homes may all have a means of disseminating your safety messages.

Movie Theater Advertising

If the movie theaters in your area show advertising slides on the movie screen before the picture, you have yet another opportunity to deliver fire prevention messages throughout the year.

Many businesses advertise this way. The theaters project 35 mm slides on the screen for about twenty minutes before showing the feature film. The system employs a continuous carousel of slides. Each slide is shown for approximately fifteen seconds.

Create a few fire safety slides. A cost-efficient way to do this is to use a computer program, then have a local photo shop create slides from your disk. Since most projection systems will crop your art in some way, have the photo shop reproduce the art appropriately, keeping the message away from the borders of the slide.

Then, set up an appointment with the theater manager to present your slides. The first theater manager we approached eagerly agreed to display them at no charge. Within a few days, our slides were shown prior to each movie in all ten theaters in the complex. You may want to present generic as well as seasonal themes. There is limited space on slides and limited time to read them, so the messages must be quick reminders only.

Chapter 14

Props, Gifts, and Books

Smoke Detector Demonstration Unit

A simple unit that demonstrates the effectiveness of smoke detectors has uses in many of the programs that we present. You can build such a unit at little cost, expending little time.

The materials you'll need are readily available at the hardware store. A simple unit consists of two pieces of plywood, eight inches square. Use these as the top and bottom of a box, connected on three sides by 24- X 8-inch pieces of acrylic. The fourth side remains open. Attach a smoke detector to the top piece of plywood, inside the box. Then place an ashtray and a lit cigarette inside the chamber. This demonstration unit will show just how little smoke it actually takes to activate a smoke detector. In just a few seconds, the detector will activate when light smoke fills the upper part of the chamber.

If you can, use a smoke detector that was involved in a house fire in your district, because it will stimulate a lot of interest among your audience. Even though it's probably discolored (good) and melted (even better), you can stress that it performed its job of warning the family. If it didn't activate, emphasize the reasons it failed. Make it clear that a detector doesn't do anyone any good if it isn't operational.

Coffee Can Fire Extinguisher

This project is suitable for most children to build on their own. Scout leaders and teachers may want to use this as a project. Keep copies of this description on hand to present to them.

The materials you'll need to create a coffee can extinguisher are as follows:

- Coffee cans with lids.
- Enough baking soda to fill the number of coffee cans you intend to make.
- A wooden dowel to form the handle, approximately one inch in diameter and five inches long.
- Wooden spools of thread.
- Two nut-and-bolt combinations, plus washers. These should be long enough to go through the spools and handle into the coffee can.
- A drill with a bit large enough to create holes for the bolts.

First, measure and mark the side of the coffee can for the two holes for the handle. Drill them about four inches apart or a little more. Then drill two holes in the dowel the same distance apart. Insert the bolts through the dowel and spools, then through the holes in the can. Attach and tighten the nuts to the bolts, using the washers. This creates a handle for the can.

Paint and label the coffee can for its new use: Coffee Can Fire Extinguisher. Then, fill the can with baking soda, and cover it with a plastic lid.

Keep the extinguisher near the stove at home, take it on camping trips, or give it as a gift to your parents or other adult.

First Aid Kit in a Can

This kit is especially suited to baby-sitter classes and scouting groups of all ages. Take one of these with you whenever you give a talk on what to do until emergency medical help arrives.

You can make an effective and inexpensive first aid kit using an empty two- or three-pound coffee can with a lid, or some other similar-size container. These kits are suitable for baby-sitting assignments, camping trips, hiking expeditions, sports activities, boats, automobiles, and elsewhere.

The contents of the can are as follows:

- One triangular bandage.
- Twenty bandage strips.
- Four 2-inch by 2-inch sterile gauze pads.
- Four 4-inch by 4-inch sterile gauze pads.
- One roll of 1-inch roller gauze.
- One roll of 2-inch roller gauze.
- One roll of cloth tape.
- Ten cotton swabs.
- One small bar of soap.
- One small hand towel.
- Five small packages of individual cleansing wipes.
- Four large safety pins.
- One pair of safety scissors.
- A pocket mask.
- Latex gloves.
- Tweezers.
- Aspirin.
- Pencil and paper.
- Several quarters taped to the inside of the lid for emergency phone calls.

Bookmarks

Bookmarks, as well as other handout items, can provide the recipients with a quick reminder of fire safety tips. Give them away to tour groups, organizations that you visit, and walk-ins to your booth at the Open House. They make great giveaways when your department is asked to send a representative to read a book to schoolchildren as part of a reading fair or some other occasion.

Bookmarks are quite inexpensive to make. You can purchase them, but the

Handout items such as bookmarks can help reinforce basic fire safety tips.

message will be generic, and the bookmark usually won't be personalized. Even if personalization were an option, it would usually make the item more expensive.

Department members and relatives may be able to donate their special talents for creating bookmarks and designing personalized messages. Elicit the help of local art students if you can, and compensate them, even if only with a T-shirt or some other memento.

To mass produce the bookmarks, contact the school system's printing office or a local printer. Our local school printing office produced bookmarks on heavy, colored paper; cut them to size; and boxed them for us. Recently we had 10,000 bookmarks made for less than $200, and they'll last for several years.

Fire Truck Bookcase

There are many different types of fire truck bookcases in public libraries across the country. Many were put together with donated materials and assembled without cost by firefighters. Children are naturally attracted to fire trucks. If these bookcases are properly outfitted with fire safety books and videos that can easily be checked out of the library, then the cause of fire safety education will be served.

One of the smaller bookcases I've seen is a one-sided, lateral depiction of a truck that sits against the wall. The bookshelves are arranged along the bed and compartment areas of the truck. You can make this type of bookcase a little larger and more three-dimensional by completing the cab and tailboard sections; i.e., by adding sides to the bookcase and painting them to look like the front and rear of the truck. This will move the bookcase away from the wall and give you more room to house books in the hosebed on top.

Of course, stand-alone units offer much more shelf space and exposure. Such a bookcase may be decorated on all sides if it stands in the open.

Fire truck bookcases are common in children's libraries across the country.

An even more elaborate option is to make the unit big enough for the children to enter on one end and sit on a padded bench while choosing the book they want.

If members of your department construct such a bookcase, try to get as much PR value out of it as possible. Contact your local broadcast and print media to make it publicly known that the department built it, donated it to the library, and stocked its shelves with safety-related books and videos. Be sure to thank all who made the display possible, whether they drove nails for it or not.

Periodic reminders to the community about the bookcase will help keep it in people's minds. Teachers and parents may then plan outings to the library to research fire safety and discuss the subject, possibly checking out a book or two at the same time. Teachers may even want to assign fire safety books as subject matter for book reports. Consider asking a school to invite one of its classes to the library so that a firefighter can read to the group. This will further stimulate their interest in the bookcase and fire safety material.

Video Library

An inexpensive way to promote fire safety education within your community is to develop a video lending library. In addition to loaning them out, you can also incorporate them in your various presentations. There are commercially available videos on fire safety for all age groups, many of them at reasonable prices. Once you've purchased a few, you can advertise your new lending library by some of the means mentioned above. Set aside an area in a fire station for people to come and browse through the video collection, not to mention any books that you may care to lend. Record the name and number of the borrower as each title goes out. This log will prove helpful when assessing the overall merits of the program.

Even a few inexpensive videos can prove financially prohibitive for some departments. Local businesses and community service clubs may help sponsor the program. In turn, your department can show its appreciation by placing a sticker on the tape, noting the donation and thanking the club or business for its gift.

"This Is Fire" Kit

One of the most versatile kits available to help spread the fire safety message is the "This Is Fire" kit, available without charge from the Federal Emergency Management Agency (FEMA). The kit contains numerous camera-ready print PSAs that can be used in local newspapers, magazines, and newsletters.

The kit also contains five radio PSA scripts, which may be photocopied and dropped off at your local radio stations. When I dropped ours off, I suddenly found myself parked behind a microphone, recording the messages myself. These announcements lend themselves to a variety of vocal interpretations, from conversational to deeply serious. If you get the opportunity to record your own PSAs, have fun. These scripts are also appropriate as voiceover copy for spots on cable or broadcast TV.

If you're looking for some safety-related material that can get you started right away and at no cost, order one of these kits from the Federal Emergency Management Agency.

Chapter 15

Games and Tests

Help the Fire Dog

This game, shown on page 166, is suitable for the very youngest grades. Its purpose is simply to help them readily distinguish between good fires and bad ones.

Fire Engine Drawing

This piece of art, depicted on page 167, is also tailored for younger children. Its purpose is simply to help them recognize the components of a fire truck—shapes that may at first seem puzzling and strange to their eyes.

Who Said It?

This game is appropriate for students in the upper elementary grades. Its objective is to help them realize how many people are involved in any given emergency situation, as well as what each might experience.

To begin, write the following categories on the board: patient, caller, emergency dispatcher, and EMS responder. Read the children quotes, such

Help the Fire Dog put out the fires

Put the X through the bad fires and circle the good ones

Games and Tests

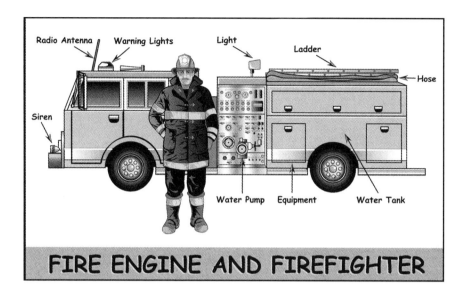

FIRE ENGINE AND FIREFIGHTER

as the following from an EMS incident, and ask them in each case who the speaker might be.

- "911. What is your emergency?"
- "Let's get this patient to the ambulance now!"
- "My daughter fell out of a tree and is unconscious!"
- "Doctor, I'm here with the patient now, and his vital signs are improving. We'll be at the hospital shortly."
- "Put the phone down, but do not hang up."
- "Should I move him into my car, where it's warmer?"
- "Can you tell me if you're allergic to any medications?"
- "Don't move the patient; you may make his injuries worse. If you have a blanket, cover him to keep him warm."
- "Let's put flares on the road around the scene to warn oncoming traffic of the crash."
- "Should I drive to the hospital myself?"
- "The ambulance is on the way as I am talking to you."
- "Should I send my friend to meet the ambulance while I stay with my father?"

- "Is anyone hurt?"
- "My name is Jeff Jones. I'm at 1234 Brooks Avenue, and my phone number is 555-6789."
- "Have someone flag down the ambulance outside of your home."
- "Please lie still while we lift you onto the stretcher."
- "We're halfway to the hospital. Can you breathe better now?"

I'll offer two suggestions on this game. First, to keep the attention of the students, try such techniques as asking the questions in different voices. Second, at the end of the game, tell the students that they're all winners.

Fire Safety Quiz

The following quiz is written for the upper elementary grades. The answers and explanations appear at the end.

FIRE SAFETY QUIZ
Choose the answer that is most correct.

1. *Most fire deaths result from...*
    - **(A)** Fire burns.
    - **(B)** Breathing too much smoke.
    - **(C)** Jumping from windows to escape.

2. *In terms of fire safety, when you go to bed at night, you should...*
    - **(A)** Close the bedroom doors.
    - **(B)** Open the bedroom doors.
    - **(C)** Leave the bedroom doors partially open.

3. *Which of the following items isn't needed for a fire to occur?*
    - **(A)** Fuel (that is, a combustible material).
    - **(B)** An ignition source (that is, a heat source).
    - **(C)** Nitrogen.

4. Smoke always contains...
   - **(A)** Carbon monoxide.
   - **(B)** Ammonia.
   - **(C)** Methane.

5. If you find yourself in a smoke-filled house, you should...
   - **(A)** Keep your head up and call for help.
   - **(B)** Keep your eyes closed and walk out.
   - **(C)** Crawl low out of the smoke.

6. If your clothes catch on fire, it is best to...
   - **(A)** Run to put out the flames.
   - **(B)** Stop, drop to the ground, and roll.
   - **(C)** Take off the burning clothes.

7. If there is fire in an oven, the best thing to do is...
   - **(A)** Turn off the heat and leave the door closed.
   - **(B)** Turn off the heat and open the door.
   - **(C)** Take the burning food outside of the home.

8. For minor burns, you should...
   - **(A)** Apply cool water to the burn.
   - **(B)** Bandage the burned area tightly.
   - **(C)** Put butter on the burn.

9. About how much time will you have to escape from a fire in your home?
   - **(A)** Less than three minutes.
   - **(B)** Eight minutes.
   - **(C)** Fourteen minutes.

10. In what room do the largest number of fires start?
    - **(A)** Attic.
    - **(B)** Basement.
    - **(C)** Kitchen.

11. *True or false: If a fire occurred while you were sleeping, the smoke would wake you up.*
    (A) True.
    (B) False.

12. *If food in a frying pan catches on fire, which is the best action to take?*
    (A) Cover the frying pan with a lid and turn off the heat to the stove.
    (B) Pour water on the frying food.
    (C) Pick up the frying pan and quickly take it outside.

13. *If an electrical cord has exposed or frayed wires, you should...*
    (A) Replace the entire cord.
    (B) Replace the frayed area only.
    (C) Cover the exposed area with a carpet.

14. *How should you react if your smoke detector goes off during the night?*
    (A) Get your valuables together.
    (B) Look in the house for a fire.
    (C) Wake up everyone and get outside.

15. *If a fire breaks out in your home, the first thing you should do is...*
    (A) Turn off the electricity to the house.
    (B) Call the fire department from your phone.
    (C) Leave the home.

16. *When do the greatest number of home fire deaths take place?*
    (A) Midnight to 4:00 a.m.
    (B) 6:00 a.m. to 10:00 a.m.
    (C) 6:00 p.m. to midnight.

Fire Safety Quiz Answers

1. **(B)** Smoke inhalation accounts for the majority of fire deaths. Develop and practice a home escape plan. The more you practice it, the better your chances of survival.
2. **(A)** Close the bedroom doors to limit the spread of heat and smoke, as well as to give you time to find another way out.
3. **(C)** Heat, fuel, and oxygen are needed for a fire to occur.
4. **(A)** Smoke contains carbon monoxide, a deadly gas.
5. **(C)** Crawl low under smoke to the nearest exit.
6. **(B)** Running or moving around will only fan the flames and increase your chance of injury. If your clothes catch fire, stop where you are, drop to the ground, cover your face with your hands, and roll over and over to smother the flames.
7. **(A)** Turn off the heat to the oven and keep the door closed. This will limit the oxygen being supplied to the fuel.
8. **(A)** Cool a burn with cool water, running it directly over the affected area. If the burn is serious, call the appropriate number to activate the emergency medical services.
9. **(A)** In a fire, survival is measured in seconds, since fire spreads rapidly.
10. **(C)** Most fires in the United States take place in the kitchen. Never leave cooking unattended, turn pot handles toward the back of the stove, and keep a lid handy when cooking.
11. **(B)** Smoke wouldn't wake you up, it would make you sleep and distort your judgment and coordination.
12. **(A)** Carrying the pan will cause the flames to flicker and fan dangerously toward you. Water will only splatter the burning grease.
13. **(A)** Don't take chances with worn electrical cords. Replace the entire cord.
14. **(C)** Smoke detectors give you a warning of fire so that you'll have time to escape. Make it your priority to leave the building.
15. **(C)** The first thing to do is to leave the building. Once outside, you can meet your family members at a predetermined outside meeting place. Then dial 911 from a neighbor's phone.
16. **(A)** Roughly, three out of every ten deaths in residential fires occur

between the hours of midnight and 4:00 a.m., when most people are asleep. This underscores the importance of installing smoke detectors to give you advance warning of a fire.

Adult Quiz

Here's a ten-question quiz that you can administer to the adult audiences at your safety seminars. The tone of some of the false answers shouldn't be considered unnecessarily frivolous or insincere, for the ultimate goal, as always, is to get people to think more about fire safety in their daily lives. Sometimes a bit of humor helps someone remember a lesson better. Again, the answers and explanations follow.

ADULT FIRE SAFETY QUIZ
Choose the answer that is most correct.

1. *Why should you have a smoke detector in your house or apartment?*
   (A) Because the law requires every home to have one.
   (B) To provide an early warning of fire.
   (C) To let you know that it's time to clean your oven.
   (D) All of the above.

2. *A home should have at least one smoke detector on each floor, especially outside the...*
   (A) Bedrooms.
   (B) Living room.
   (C) Kitchen.
   (D) Bathroom.

3. *The maintenance on a battery-operated smoke detector includes...*
   (A) Painting the detector to match your room.
   (B) Replacing the detector every three years.

(C) Removing the batteries if the alarm goes off while you're cooking.

(D) Changing the batteries at least once a year and occasionally dusting the inside of the unit.

4. *How can you best prevent cigarettes from setting a fire in your home?*

(A) Never smoke in bed or when you're sleepy.

(B) Extinguish all cigarettes with fire extinguishers.

(C) Keep smoke detectors near the smoking areas of your home.

(D) Quit smoking.

5. *What is the number one cause of home fires that result in fatalities?*

(A) Lightning.

(B) Electrical equipment.

(C) Smoking materials.

(D) Mexican food.

6. *True or false: Most fire deaths result from burns.*

(A) True.

(B) False.

7. *If your clothing were to catch fire, the best thing to do would be to...*

(A) Run to the bathtub and wet yourself down.

(B) Sit still and yell for help.

(C) Stop where you are, drop to the ground, and roll back and forth to smother the flames.

(D) Take off your clothes.

8. *If you receive a mild burn while cooking, you should...*

(A) Smear butter on it.

(B) Run cool water over it.

(C) Apply a molasses ointment to it.

(D) Order takeout next time.

9. *What should you do if food in the frying pan catches fire?*
   (A) Carry the pan to the sink and run water on it.
   (B) Smother the flames with the lid and turn off the heat to the burner.
   (C) Pour water from a glass onto the fire.
   (D) Sprinkle baking soda on nearby objects that may catch fire and let the flames die down by themselves.

10. *When escaping a room filled with smoke, the safest air to breathe is located:*
    (A) Near the ceiling.
    (B) Near the floor.
    (C) At eye level.
    (D) In the nearest oxygen tank.

ADULT FIRE SAFETY QUIZ ANSWERS

1. **(B)** The purpose of a smoke detector is to provide early warning of smoke given off by a fire.
2. **(A)** The most important place to put working smoke detectors is outside of the bedrooms.
3. **(D)** Changing the batteries in the smoke detector annually and occasionally cleaning dust from inside the unit will help keep your smoke detectors in good working order.
4. **(A)** Never smoke when you're sleepy.
5. **(C)** Smoking is the number one cause of home fire deaths.
6. **(B)** Smoke is the number one killer in fires, not burns.
7. **(C)** Stop, drop, and roll to extinguish the flames if your clothing catches fire.
8. **(B)** Cool a burn with plenty of cool water. Seek medical attention if the burn is serious.
9. **(B)** Smother the fire by putting your hand in an oven mitt and sliding the cover over the burning pot. Then turn off the heat to the burner.
10. **(B)** Since heat and fire gases rise, the safest air to breathe is near the floor. The best way to escape is to crawl low, under the smoke.

Chapter 16

Sample Handouts

Home Hazards Brochure

The following is the text of a brochure that my department frequently gives to the public at Open Houses, after presentations, and on other occasions. The brochure consists of a 8 ½ X 11-inch sheet of blue paper turned lengthwise and folded twice, triptych-style. A bit of clip art complements the copy.

IDENTIFYING FIRE HAZARDS IN YOUR HOME
How Fire-Safe Is Your Home?

Make fire safety everyone's business by involving the entire family in a fire safety inspection. Here's a comprehensive checklist to use as a guide.

Check for fire hazards in your home. Correct any problems NOW.

Kitchen

__ Matches stored out of the reach of children.
__ No overloaded outlets or extension cords.
__ No worn or frayed electrical cords.
__ No curtains or towel racks close to the range.
__ Flammable liquids (cleaning fluids, contact adhesives, etc.)

and aerosols stored away from the range and other heat sources. Remember, even a pilot light can set vapors on fire.
___ No attractive or frequently used items stored above the range where someone could get burned reaching for them (especially small children in search of cookies or other goodies).

Living Room, Family Room, Den, Bedrooms

___ Matches and lighters stored out of the reach of children.
___ Use only large ashtrays. (Small ashtrays are too dangerous.)
___ Empty ashtrays frequently (when all signs of heat and burning are gone).
___ Fireplace kept screened and cleaned regularly.
___ No overloaded outlets or extension cords.
___ Replace worn or frayed electrical cords of any type.
___ No electrical cords run under rugs or carpets, or looped over nails or other sharp objects that could cause them to fray.
___ All portable heating equipment kept at least three feet away from curtains, furniture, and papers.

Basement, Garage, and Storage Areas

___ No newspapers or other rubbish stored near the furnace, water heater, or other heat source.
___ No oily, greasy rags stored, except when kept in labeled and sealed nonglass containers (preferably metal).
___ No gasoline stored in the house or basement. Gasoline should be stored away from the house in an outbuilding, and only in safety cans that have flame arresters and pressure-relief valves.
___ No flammable liquids stored near the workbench or pilot light, or in anything other than labeled, sealed metal containers. (This includes varnish, paint remover, paint thinner, contactadhesives, cleaning fluids, etc.)
___ No overloaded outlets or extension cords.
___ All fuses are of the correct size.

Dangerous Actions

Do you allow unsafe behavior? These guidelines may help your family become more safety conscious.

___ Wear close-fitting sleeves while cooking. (No loose sleeves, shirts, blouses, or skirts that may catch fire.)
___ Never leave cooking food unattended.

Sample Handouts

___ Never play with matches or lighters.

___ Never use gasoline to start a fire in the grill or add lighter fluid to a fire that's already burning.

___ Never smoke in bed. Never smoke in a chair or on the sofa when you're drinking, on medication, or sleepy.

___ Never spray aerosols while smoking or when near a space heater, range, or other ignition source.

___ No smoking while using cleaning fluid, paint thinner, or other flammable liquid.

___ Never use a cigarette lighter after spilling flammable liquid on your hands or clothing.

179

- Never reach over a stove when it's lit or climb onto a stove to reach something stored above it.
- Never lean against a stove for warmth or stand too close to a heater or fireplace.
- Never let a small child blow out a match.
- Never use a lighted match, lighter, or candle to illuminate a closet or other dark area.

Safety Measures

Do you have these safety items in your home?

- Working smoke detectors on all levels and outside the sleeping areas (on each floor or wing, in the basement, etc.). Have you tested the detector in the past month? Have you changed the batteries within the past year?
- Is there an ABC fire extinguisher in the kitchen and the workshop?
- Do you have an emergency exit plan, including two ways out of each room and a designated meeting place outdoors?

Home Escape Plan Handout

The following is a checklist you can give your audience members to help them create their own home escape plan. Like the home hazards checklist, it is suitable for an 8 ½ X 11-inch sheet of paper folded into a brochure.

HOME ESCAPE PLAN

Preparing Your Escape Plan

- Make a floor plan of your house.
- Draw all doors, windows, and stairways.
- From each bedroom, decide which routes would be best for a quick escape. Choose at least two means of escape.
- Draw the normal exit routes on your plan with solid black lines.
- Draw the emergency exit routes on your plan with dotted colored lines.
- Indicate the location of the smoke detectors. Ensure that they're operational, that they're placed properly, and that there is a sufficient number of them.

Sample Handouts

___ Designate an area outside the home as a safe meeting place for all residents and guests.

Practice Your Escape Plan

___ Begin the drill with each family member in his or her bedroom. The bedroom doors should be closed.
___ Sound the alarm.
___ Have everyone get down and crawl along the floor.
___ Each person should check his or her bedroom door by feeling to see whether it's hot.
___ If the door is hot or if smoke is coming in, use the emergency route, which is probably through the window.
___ If the door isn't hot and if no smoke is evident, open the door slowly. Be ready to shut it again quickly.
___ Make certain beforehand that the windows will open easily.
___ Be ready to help infants or the elderly.
___ Everyone should meet at the designated meeting place.
___ Make sure everyone is accounted for.
___ Practice both normal and emergency-route scenarios.

If a Fire Does Break Out

Try not to panic. Follow the plan as you've practiced it.

___ Don't bother getting dressed or gathering valuables.
___ Check your bedroom door before opening it.
___ Wake up the household immediately.
___ Get out of the house and go straight to the designated meeting place.
___ Never go back inside a burning building.

Baby-Sitter's Handout

Use the following copy to create a handout to give to baby-sitters and other guardians at the end of the baby-sitting presentation.

BABY-SITTING IS A BIG JOB

Keeping children SAFE is the number one responsibility of a baby-sitter. Being prepared is essential—there is no time to prepare once an emergency occurs. One of the many emergency situations that you could face is fire.

- Keep matches and lighters where children can't reach them.
- Keep portable heaters away from play areas and other combustibles.
- Turn pot handles toward the back of the stove so children cannot pull on them.
- Do not cook while holding a child.
- Do not wear tight-fitting sleeves when cooking.
- When cooking, keep the lid of the pan nearby to smother the flames if a fire starts in the pan.
- Always check the temperature of the food before giving it to a child.
- Show children how to stop, drop, and roll if their clothing catches fire.
- Put cool water on a burn. If the child gets burned or receives some other serious injury, call 911 immediately and the child's parents thereafter.
- Ask the parents about a home escape plan before they go out.
- Know the sound of the smoke detector and what to do if it goes off.
- If you smell smoke, see flames, or hear a smoke detector go off, get everyone outside.
- If you cannot get to the children, notify 911 and tell the dispatcher that children are still inside the house. Tell this to the first-arriving firefighters also.

Career Day Handout

My department uses the following text in handout form to give to interested students on career day. For this brochure, we use an 11 X 17-inch sheet turned lengthwise and folded thrice to create four panels.

THE FIREFIGHTER NOW
The Life and Work of a Contemporary Firefighter

The Firefighter

The ancient Greeks considered fire to be one of the four elements. Throughout history, fire has been one of our greatest natural allies, used for cooking, heating, and making tools, as well as for industrial and military purposes. It has also been one of our greatest natural enemies. In fighting it, we have relied on everything from leather buckets to horse-drawn pumpers to modern computer-operated, high-pressure deluge apparatus.

Just as society has become more complex, so too has the fire danger and our means of combating it. Aerial ladders, adjustable nozzles, amphibious apparatus, foam, and powdered chemicals are just a few of the newer developments. In little more than a century, firefighting has passed from being an occasional voluntary community effort to a highly scientific profession.

Still, none of our knowledge, methods, and equipment have replaced the key element in preventing, confining, and extinguishing fires—the firefighter. Nothing has been developed that'll take the place of the person who meets the fire head-on and defeats it. A firefighter performs one of the most dangerous jobs in the world.

Firefighters have been glamorized, satirized, and sometimes politicized, but they remain among the mainstays of modern civilization. In an age where many jobs require a compromise of integrity, the firefighter remains someone who can walk tall. Those in the fire service can always be proud of what they do with their hands, heart, back, and brain, often at the risk of their own life. This blending of tradition, skill, knowledge, bravery, and technical know-how is sure to continue. The fire service is a proud profession, honoring its past and looking forward to its future.

What Does a Firefighter Do?

Whether employed by a large city, a county, a suburb, a company, or an airport, or even serving as a volunteer, a firefighter is someone with a mission as sacred as can be, for a firefighter's task is to save lives and property.

This is done by preventing fires whenever possible. Prevention is a science in itself. Planning the strategy and tactics just in case the preventive efforts fail, as well as actually going into battle against the flames, is a larger part of the job. Whether the fire is in somebody's kitchen, a public building, a factory, or a forest, the mission remains one of the most challenging and demanding tasks that a human being can undertake.

Firefighters can either be specialists or generalists. The generalists, like the infantry in combat, are the heart and soul of firefighting. The modern firefighter needs to know more than those of days gone by, even those of the recent past. They must know more about personal protection and how to deal with a combination of new dangers, such as toxic gases, explosives, and hazardous materials. The demands on today's firefighter are great, and the need for continuous learning and the development of new skills is ever increasing.

After completing time in the general fire service, one may seek or be assigned to specialized duty, such as inspection, investigation, public relations, communications, special rescue units, or paramedic duty. Promotional opportunities are available within both the general and specialized fields. All of these endeavors require study, training, practice, and tremendous dedication.

It is a tradition that firefighters be called for all kinds of emergencies, not just those involving fire. Firefighters may respond to freeway accidents, train derailments, and medical emergencies—in fact, virtually anything that no other service can or wants to tackle falls within the domain of the fire service. That's part of the appeal of the job. On any given day, a firefighter can answer a call and witness situations that few others will ever know or experience.

Who Can Be a Firefighter?

There was a time when anyone willing to go to a rural fire or who had the right connections in a city could be a firefighter. That time has passed.

It is an accepted fact that not everyone can or should be a firefighter. The National Fire Protection Association has developed Standard 1001, which specifies personal standards and practical performance standards for firefighters at each level of the profession.

The applicant doesn't have to be superhuman, but he should be fit, agile, and able to work hard under stress for long periods of time. Any serious health disorder or physical handicap will probably disqualify a firefighter candidate. Great size isn't necessary, but the ability to withstand extremes of heat and cold; to exert oneself to the maximum; and to carry considerable weight are essential.

The qualifications also usually include being a U.S. citizen and passing both a written examination and rigorous physical testing that measures an applicant's strength, stamina, and agility.

Many fire departments require an applicant to submit proof of EMT licensure and an acceptable driving record.

It is also important that the candidate be mentally and

Sample Handouts

emotionally sound. There is both boredom and great tension in firefighting. There is a need to be cool and cautious without being finicky or hesitant. Sometimes a firefighter has to make life-and-death decisions almost instantly. Not everyone can handle the demands of the job, and each fire department must ultimately determine who can and who can't make the grade.

Some mechanical aptitude and manual dexterity are necessary. Most firefighting skills can be learned and mastered with patience and practice, but the person who is a little handy has an edge.

At present, it is necessary to be at least eighteen years of age to become a firefighter. Persons over forty-five aren't encouraged to enter the field.

Many fire departments have a residency requirement, mandating that employees live within their jurisdiction or move to it within a specified time.

Good character, though hard to define, is another trait that firefighters must have. Firefighters must be trustworthy. As team players, they must have self-discipline and learn to function under a command system. Carelessness, impulsiveness, or too great an affection for heroics can endanger lives and property. A firefighter often must work quickly, but patience is as important as celerity.

There is one more requirement, and it is one that is becoming more and more indispensable. A firefighter should be willing to train, then train some more, then go back and train even more. In between these periods, the ideal firefighter will practice or review what he or she has learned. Many of the necessary skills, such as tying knots, won't be used very often, but the firefighter must be proficient in them, anyway. When those skills are required, it may be that someone's life depends on them being performed correctly.

What Role Do Firefighters Play in Delivering Emergency Medical Services?

Firefighters play an essential role in providing emergency medical services (EMS) to the citizens they serve. Today's firefighters don't stop after rescuing someone from a burning building. They follow up with prehospital emergency care, administered by state-licensed emergency medical technicians (EMTs).

Although licensed by the state Department of Public Health, EMT training programs are available through colleges, some teaching hospitals, and some private companies, including ambulance services.

In addition to their firefighting duties, EMTs perform a wide repertoire of emergency functions. Whether responding to vehicle accidents or home emergencies, EMTs must be prepared to deliver their lifesaving skills under a variety of circumstances, often less than optimal.

Some fire departments only act as a first responder, allowing an ambulance company to provide the transportation to a medical facility. Other fire departments provide transportation services as well, taking patients to the emergency room.

To maintain their EMT licensure, firefighters must complete frequent continuing education courses as required by the state.

Training to Be a Firefighter

In the past, firefighters learned their trade on the job. Someone showed them where the engine was and told them to be sure to be on it when it rolled.

In today's world, that kind of approach won't do anymore. Most city firefighters undergo many weeks of full-time training before receiving their first assignment. They are then trained as probationary firefighters, taking part in training and drills with other members of their companies. The probie who doesn't realize that learning never ends probably won't be on the job for very long.

For their own safety and the welfare of others, firefighters must have considerable knowledge of fire science. They must also know something about the law; understand different types construction; know the best ways to obtain and apply water; master the art of climbing and working under hazardous conditions; understand how to prevent explosions; know how to find and rescue victims; and even be able to provide competent medical treatment to patients until more qualified medical personnel take over.

Where Can You Learn More About Being a Firefighter?

The best answer to that question is, "Go to the fire station." Go in person or send for the requirements of any department you're interested in joining. Requirements change frequently, so you would be well advised to check frequently.

Some firefighters aren't paid anything at all. Others are paid per run. Full-time firefighters work as little as forty hours a week or as much as fifty-six hours, and sometimes more. Shifts of twenty-four hours are common, but there's usually ample time off in between shifts. The pay scales vary greatly, depending on the community. Officers and specialists can earn considerably more than line firefighters.

The fringe benefits are generally good, and there are special provisions for the disabilities that firefighters may suffer. The retirement plans are usually the best available.

To find out the details, check with the departments that interest you. Many candidates tour a number of departments to find the one with the best opportunities. Often a candidate must wait some time before getting an appointment. That time can be spent in study or in physical conditioning. Among firefighters, you'll find distance runners, weight lifters, handball champions, and a lot of ball players. It's a job where there's no such thing as being too healthy.

So What Now?

This brochure has given you a brief look at the fire service and how you might fit into it. If you qualify and have the motivation to take on the training, risks, labor, and all else, we encourage you to give it a try.

It'll require a lot of work, worry, and effort, but many people continue to find that the fire service offers a satisfying way of life. This profession may demand a lot, but it also guarantees one thing: You'll never doubt whether your work is important or not.

Medical Emergencies Handout

As mentioned in Chapter 7 and elsewhere, a handout covering emergency medical care can be of great benefit to those who may need to care for a patient before help arrives. Following is a fairly substantial amount of material covering layman responses to a range of common medical circumstances. Although photocopying and stapling this information is certainly an inexpensive option, your department may wish to use it as the basis for a more professional-looking brochure or booklet.

EMERGENCY ACTION: WHAT TO DO UNTIL HELP ARRIVES

The objective of this class is to increase your personal knowledge. You can acquire the practical skills by enrolling in an advanced class.

In the event of a medical emergency, notify EMS immediately by dialing 911. The emergency number may be different in your area, so be familiar with it beforehand. Give the dispatcher as much information as possible.

Heart Attack

- Heart attack is the number one killer in the United States. Be familiar with its signs and symptoms: major chest pain, numbness or tingling in the upper arms or neck, diaphoresis (cool, clammy sweat), shortness of breath. Usually the patient will deny these symptoms. Call for help immediately.
- If the patient stops breathing before medical help arrives, administer cardiopulmonary resuscitation (CPR).

Unconscious Victim Priorities

Not all unconscious victims need the assistance of rescuers. You can assess an unconscious victim using the ABC (airway, breathing, circulation) method. The results of this assessment will indicate to the rescuer whether lifesaving protocols are needed.

- On finding an unconscious victim, first ensure that it is safe to approach the scene where he or she is lying.
- Determine whether the patient is truly unconscious by shaking him and shouting.
- Roll the victim onto his back as a unit. Support the head during this maneuver, and ensure that it doesn't hit the ground.
- Open the airway using the head-tilt method. The head should be resting on the ground, on the same plane as the shoulders, and it should be in line with the body, not rolled to one side. The neck should be extended.
- Assess the breathing by looking, listening, and feeling for an air exchange from the nose and mouth for approximately five seconds. If there is no sign of breathing, pinch the nose and breathe twice into the victim's mouth.
- Check the circulation by feeling for the carotid pulse in the neck for five to ten seconds.
- Consult current manuals for a proper depiction of these skills.

Obstructed Airway

- Choking kills more than 3,000 people annually in the United States alone. Choking deaths are sometimes called "cafe coronaries," because a choking victim can appear to be having a heart attack. However, the choking victim's airway is closed by the obstruction. A heart attack victim's airway is open.

- Remember the universal distress signal of an obstructed airway: The person will be holding his throat with his hands.
- Use the correct hand position to deliver abdominal thrusts to relieve the obstruction.
- You can administer the abdominal thrusts yourself if you are alone and choking.

Poisoning and Drug Overdose

- Always keep the phone number of the local Poison Control Center posted near the telephone.
- Some common signs of poisoning include finding open bottles of medication or other substances, such as cleaning products and kitchen supplies. A half-eaten plant may also be a sign of poisoning. The symptoms of poisoning include abdominal pain, nausea, vomiting, diarrhea, convulsions, difficulty in breathing, and sweating.
- In the case of an obvious poisoning or drug overdose, call 911 immediately. If a minor is involved, also call the parents.
- If the incident doesn't appear life-threatening or you aren't sure poisoning has occurred, call the Poison Control Center. As many as 85 percent of all calls are handled in the home under the direction of the Poison Control Center.
- Attempt to determine the victim's age and weight, as well as his level of consciousness, the type of poison, etc.
- Dilute the poison or induce vomiting? Use activated charcoal to absorb the poison in the stomach. Use syrup of ipecac to induce vomiting. If the person is unconscious, do not attempt either one. Neither should you attempt these procedures except under the advice of the Poison Control Center.

External Bleeding

- The standard first aid care for minor wounds is to clean and dress the wound.
- If the bleeding is heavy, elevate the wound site above the level of the heart so that gravity will help stop the bleeding. Apply pressure to the wound site with the cleanest item available.

Nosebleeds

- Make the patient calm and comfortable.

- Keep yourself calm.
- Have the patient breathe through his mouth.
- Pinch the patient's nostrils together until the bleeding stops.

Amputation
- Find the amputated limb.
- Keep the limb cool, not frozen.
- As much as possible, keep the amputated limb out of the victim's view.
- Try to control the bleeding. Use latex gloves, and avoid coming in contact with the patient's body fluids.

Burns
- Most first- and second-degree burns will benefit greatly if cool water is applied for five minutes.
- Apply cool water to the affected area.
- Wrap serious burns in dry, sterile dressing.
- Do not apply butter, margarine, ointments, salves, elixirs, or the like.
- Do not use ice. It could damage skin tissue.
- Avoid home remedies.
- Call EMS.

Smoke Inhalation
- Have working smoke detectors in your home.
- Remember to stop, drop, and roll if your clothing catches fire.
- Crawl low under smoke.

Frostbite
- Frostbite must be treated by professional medical personnel at once.
- The frostbitten area will look waxen and discolored.
- Remove the patient from the cold. Rewarm him with blankets and heaters. You may immerse the affected area in warm water.
- Do not rub the affected area to generate warmth.
- Rewarm the person slowly.
- Take the patient to the hospital.

Heat Emergencies
- This category of emergencies includes, in ascending order of severity, heat cramps, heat exhaustion, and heatstroke.

- If the patient is sweating, move him or her to a cooler area, and give him something to drink if he can swallow.
- A victim who isn't sweating but who has very hot skin is suffering from heat stroke, which is a true medical emergency. Cool the patient immediately. Call EMS immediately.

Drowning

- Prevention is the key. Plan ahead.
- Never swim alone.
- Have a telephone nearby.
- Keep rescue tools nearby, such as life rings, long-handled sticks, and other items that can be used to pull the victim out of the water. Always attempt to use the rings and tools to remove the victim so that you don't have to go into the water yourself.
- No alcohol, gum chewing, or eating while in the water.
- Keep watch on the weather conditions.
- In the event of an accident, perform the ABC assessment: airway, breathing, circulation. Call for help immediately.

Head Injury

- When an accident occurs and the head is involved, even if there is no outward sign of injury, assume there to be a closed head injury.
- Seek medical treatment. If the victim refuses medical treatment, closely monitor him by assessing his pupil reactivity and level of consciousness for a minimum of four hours. That is the time it would take for the signs and symptoms of a head injury to manifest themselves.

Splinting

- To splint a fracture means to immobilize the injury site both above and below the point of the break.
- If splinting is necessary, you should usually splint the injured part in the position in which it is lying. "Splint where it lies" is the rule of thumb to follow.
- If you have called EMS and help will arrive shortly, and if the patient isn't in a great deal of pain, you can leave the injured area alone. Right after a fracture occurs, the body instinctively finds the position that is most comfortable for it, and it will attempt to maintain that position. Help the victim find and maintain that position until EMS personnel arrive.

Spinal Injury
- If you suspect spinal injury, do not move the patient unless there is a further threat to his safety, such as if he is in a burning car.

Seizure
- Although it sounds fairly rare, seizure is a common medical emergency.
- A commonly held belief is that the victim will swallow his tongue unless some foreign object is jammed into his mouth. This is wrong. Do not jam items down the throat of someone experiencing a seizure. Doing so only risks choking him.
- Try to keep the patient from hurting himself. Move furnishings and other items that might cause injuries out of the way.
- Calmly reorient the patient if the seizure passes quickly.
- Most seizures come and go fairly quickly. If the seizures are of long duration or continual, or if the patient has no history of seizure, or if the family requests it, call 911. Generally, call if any seizure continues beyond thirty seconds, but try to alert a family member, if possible, before calling 911.
- If the victim has a history of seizures, the patient or his family may want to wait to see whether the seizure passes before calling EMS. The family, if available, will probably know whether to call EMS. In many cases, family members record the type and duration of the seizure for review by the patient's doctor.
- Protect the victim's privacy to shield him from embarrassment.

Fire Extinguisher Handout

No matter how much emphasis you place on hands-on training in the proper use of fire extinguishers, it's always important to have a handout ready to reinforce the instruction you've given.

HOW TO USE A FIRE EXTINGUISHER

Extinguisher Markings
- Class A: ordinary combustibles, such as wood, paper,

How to Use a Fire Extinguisher

Extinguisher Markings

  • **Class A**: Ordinary combustibles, such as wood, paper, cloth, rubber, many plastics, and other common materials that burn easily.

  • **Class B**: Flammable liquids, such as gasoline, oil, grease, tar, oil-based paint, and lacquer.

  • **Class C**: Energized electrical equipment, such as fuse boxes, meters, and appliances still plugged into receptacles.

 • **Class D**: Combustible metals, such as magnesium and sodium.

Before Using an Extingusher...

- Make sure everyone has left the building
- Call the fire department.
- Inform someone outside of your intention to fight the fire. Have that person tell the first-arriving fire company that you're inside the building.

Fighting the Fire

To fight a fire, remember the acronym *PASS*:

- **P**ull the pin or latch.
- **A**im the nozzle at the base of the fire.
- **S**queeze the handle.
- **S**weep the spray from side to side at the base of the flames.

Remember...

- Make sure that any fire extinguishers you use are listed or approved by an independent testing laboratory, such as Underwriters Laboratories (UL).
- An extinguisher must be fully charged before use, and it must either be recharged or replaced after each use or when the gauge indicates that the unit needs recharging.
- Extinguishers should be installed in plain view, near an escape route from the building. They should be close to, but not in, likely areas of fire, such as shops, kitchens, and garages.
- Most portable extinguishers work according to these directions, but some may not. Read and follow the directions on your extinguisher.

cloth, rubber, many plastics, and other common materials that burn easily.
- Class B: flammable liquids, such as gasoline, oil, grease, tar, oil-based paint, and lacquer.
- Class C: energized electrical equipment, such as fuse boxes, meters, and appliances still plugged into receptacles.

Before Using an Extinguisher...
- Make sure everyone has left the building
- Call the fire department.
- Inform someone outside of your intention to fight the fire. Have that person tell the first-arriving fire company that you're inside the building.

Fighting the Fire
To fight a fire, remember the acronym PASS:

- Pull the pin or latch.
- Aim the nozzle at the base of the fire.
- Squeeze the handle.
- Sweep the spray from side to side at the base of the flames.

Remember...
- Make sure that any fire extinguishers you use are listed or approved by an independent testing laboratory, such as Underwriters Laboratories (UL).
- An extinguisher must be fully charged before use, and it must either be recharged or replaced after each use or when the gauge indicates that the unit needs recharging.
- Extinguishers should be installed in plain view, near an escape route from the building. They should be close to, but not in, likely areas of fire, such as shops, kitchens, and garages.
- Most portable extinguishers work according to these directions, but some may not. Read and follow the directions on your extinguisher.

Open House Comment Sheet

Finally, you'll want some feedback from the public on your Open House. The following is based on the comment sheet that my department hands out to those who attend the event. You should elicit feedback from those who planned and worked that day, too.

Creating a Fire-Safe Community: A Guide for Fire Safety Educators

OPEN HOUSE COMMENT SHEET

Near the end of our day, or at any time you have a moment, please feel free to comment on our event and make suggestions as to how it can be improved next year. If you need more time, you're welcome to take this sheet home and send it back to us. Your time and efforts are greatly appreciated.

Event that could be improved: _____

How it could be improved: _____
_____
_____

Something that could be added next year: _____
_____
_____

Something that could be dropped next year: _____
_____
_____

Additional comments: _____
_____
_____
_____
_____

Name (optional): _____

Would you be interested in helping us next year, either in organizing, planning, or conducting the event? If so, please supply your name and contact information. _____
_____
_____
_____